ROUTLEDGE LIBRARY EDITIONS:
THE ECONOMICS AND POLITICS OF OIL
AND GAS

Volume 2

OPEC: TWENTY YEARS AND BEYOND

OPEC: TWENTY YEARS AND BEYOND

Edited by
RAGAEI EL MALLAKH

Routledge
Taylor & Francis Group
LONDON AND NEW YORK

First published in 1982 in Great Britain by Croom Helm Ltd
First American edition published by Westview Press, Inc., 1982

This edition first published in 2016
by Routledge
2 Park Square, Milton Park, Abingdon, Oxon OX14 4RN

and by Routledge
711 Third Avenue, New York, NY 10017

Routledge is an imprint of the Taylor & Francis Group, an informa business

British Library Cataloguing in Publication Data
A catalogue record for this book is available from the British Library

ISBN: 978-1-138-64127-3 (Set)
ISBN: 978-1-315-62232-3 (Set) (ebk)
ISBN: 978-1-138-64287-4 (Volume 2) (hbk)
ISBN: 978-1-138-64289-8 (Volume 2) (pbk)
ISBN: 978-1-315-62967-4 (Volume 2) (ebk)

Publisher's Note
The publisher has gone to great lengths to ensure the quality of this reprint but points out that some imperfections in the original copies may be apparent.

OPEC: Twenty Years and Beyond

edited by Ragaei El Mallakh

Westview Press • Boulder, Colorado
Croom Helm • London, England

This is a Westview reprint edition, manufactured on our own premises using equipment and methods that allow us to keep even specialized books in stock. It is printed on acid-free paper and bound in softcovers that carry the highest rating of NASTA in consultation with the AAP and the BMI.

This volume is included in Westview's Special Studies in International Economics and Business.

Published in 1982 in the United States of America by
Westview Press, Inc.
5500 Central Avenue
Boulder, Colorado 80301
Frederick A. Praeger, Publisher

Published in 1982 in Great Britain by
Croom Helm Ltd
2-10 St Johns Road
London SW11

Library of Congress Catalog Card Number: 81-7531
ISBN (U.S.): 0-86531-163-3
ISBN (U.K.): 0-7099-0904-7

Printed and bound in the United States of America

10 9 8 7 6 5 4 3 2

To my wife, Dorothea,
for her inspiration, encouragement, and support
and in recognition of her contribution to the field of energy,
in particular her editorship of the *Journal of Energy and Development*

CONTENTS

TABLES AND FIGURES

Tables

Figures

ABBREVIATIONS

ADNOC	Abu Dhabi National Oil Company
ATK	aviation turbine kerosine
bb/yr	billion barrels per year
b/d	barrels per day
BOP	balance of payments
BPY	*Balance-of-Payments Yearbook*
CEMA	Council for Mutual Economic Assistance
cf/d	cubic feet per day
CIA	Central Intelligence Agency (U.S.)
CIEC	Conference on International Economic Cooperation
c.i.f	cost, insurance, and freight
DAC	Development Assistance Committee (OECD)
DOT	*Direction of Trade*
ECG	Energy Coordinating Group (OECD)
EEC	European Economic Community
f.o.b.	free on board
GATT	General Agreement on Tariffs and Trade
GDB	Government Domestic Currency Balance
GDP	gross domestic product
GFB	Government Foreign Balance
GNP	gross national product
IEA	International Energy Agency (OECD)
IFAD	International Fund for Agricultural Development
IFS	*International Financial Statistics*
IMF	International Monetary Fund
KNPC	Kuwait National Petroleum Company
LDC	less-developed country
LNG	liquefied natural gas
LPG	liquefied petroleum gas
NATO	North Atlantic Treaty Organization
NIOC	National Iranian Oil Company
NNOC	Nigerian National Oil Corporation
NOC	national oil company
OAPEC	Organization of the Arab Petroleum Exporting Countries
OECD	Organization for Economic Cooperation and Development

OLADE	Latin American Energy Organization
OPEC	Organization of the Petroleum Exporting Countries
PDV	present discounted value
PEMEX	Petróleos Mexicános
PRC	People's Republic of China
QGPC	Qatar General Petroleum Corporation
SDRs	special drawing rights (IMF)
TER	total energy requirement
UAE	United Arab Emirates
UNCTAD	United Nations Conference on Trade and Development
UNDP	United Nations Development Fund
WAES	Workshop on Alternative Energy Strategies

INTRODUCTION

Ragaei El Mallakh

There is rarely anything lackluster in the responses to actions of the Organization of the Petroleum Exporting Countries (OPEC): they tend to be either emphatically "for" or "against." Similarly, most writing on the organization and its member countries tends to be either a diatribe or an apologia. Such polarization can lead to distortions in evaluating the objectives of OPEC, its methods, and the extent of its impact in the global energy and economic scenes.

The chapters in this volume have been selected to hew to a middle path. While the individual authors present specific viewpoints, some semblance of balance has been sought in offering both producer and consumer positions. The contributors range from academicians; representatives of OPEC, the energy industry, and international and multinational agencies; to non-OPEC producers and spokesmen for the importing nations. In a number of instances, the authors present their own opinions rather than official positions. Readers will be struck by the differing analyses and forecasts advanced, but the diversity should not be too surprising given the multifaceted nature of world energy supply and demand.

The idea of publishing this book evolved because 1980 marked the twentieth anniversary of the founding of OPEC and because the response to a special issue on OPEC of the *Journal of Energy and Development* had been exceptional. Some of the chapters in this volume were published originally in that special Spring 1979 issue of the *Journal.* Those pieces retain their validity because of the trends detected and the policies outlined. They also offer some means of gauging the accuracy of projections and of indicating the degree of flexibility – some would say instability – of the international energy picture with its sweeping impact on the world economy. Six chapters were written specifically for this book. In selecting the contributions included here, the goal has been not only to trace the evolution of OPEC and to assess its present status, but also to discern where it is going.

The article by OPEC's Secretary General René Ortiz offers the underlying rationale for the Organization's policies and spells out its priorities. The internationally recognized petroleum specialist J. E. Hartshorn asks the usually unasked question of whether, in the absence of additional political crises, OPEC will face difficulties in increasing the real prices of oil in the 1980s, as it did in the mid-1970s. Robert Pindyck has looked at OPEC oil pricing as premised upon economic maximization *à la* cartel behavior, whereon a reasonable basis for explaining and forecasting the price of oil might be provided. Although written before the late 1979-1980 round of price leapfrogging, his conclusion that oil will become increasingly expensive in coming years would appear on target.

During the First Arab Energy Conference, sponsored by the Organization of the Arab Petroleum Exporting Countries (OAPEC) and held in Abu Dhabi in 1979, a number of well-researched and well-written papers were presented; one is included here ("Optimum Production and Pricing Policies") by two OPEC economists. Fadhil Al-Chalabi and Adnan Al-Janabi offer insights into the elements that are seen as determinants in output decisions and pricing policy by OPEC states. A paper delivered to the OAPEC staff by Salah El Serafy on "The Oil Price Revolution of 1973-1974" seeks to delineate the factors underlying the quantum price jump in 1973-1974 and to assess their directions for the future.

Professor Walter Mead analyzes the 1973-1974 period in crude oil pricing as well as the longer span of the 1970s, moving to examine the supply controls in the United States as a contributing factor to supply-demand imbalances with concomitant pricing implications.

René Zentner addresses two critical issues: how the energy problem is perceived by the public, particularly in the United States, and how realistic public perceptions of OPEC and the energy-related activities and policies of both government and industry are. This articulate study is instructive as to why governmental policy is difficult to fashion when the public holds certain views – correct or incorrect – on the severity and causes of the energy imbroglio.

The Dean of the College of Industrial Management at the University of Petroleum and Minerals in Saudi Arabia, Ali D. Johany, argues persuasively that the near-obsession by many who hold OPEC *per se* responsible for the onset and depth of the energy and global economic problems clouds the real issue. The transfer of ownership rights from the companies to the producing governments gave the countries economic sovereignty and hence power to influence prices. He concludes that with or without OPEC, 1973-1974 would have seen sharp price increases in oil.

Dennis O'Brien envisages an expansion in downstream operations by the major producing/exporting nations, based on a mixture of domestic economic and political considerations. He sees these expanded activities as having very definite implications on output levels in the coming decade. O'Brien's "consumer" view has a companion piece in the article by Ali

Jaidah, now Managing Director of Qatar's national oil company and former Secretary General of OPEC. Jaidah's study draws heavily on his opening remarks to the seminar on downstream operations held by OPEC in Vienna during October 1978. It authoritatively traces the *raison d'être* behind OPEC's stated objective of diversifying within the whole range of oil industry operations and away from the role of crude oil suppliers only.

Much has been written of the hardships caused to the developing bloc by the massive oil price increases that began in 1973-1974 and were repeated at the close of the decade. OPEC's response to its fellow Third World states is embodied in the OPEC Special Fund (renamed in 1980 the OPEC Fund for International Development), whose Director, Ibrahim Shihata, has outlined the perspectives of this relatively new aid agency and its processes. Despite the existence of the Fund, the depth of the economic morass facing much of the developing bloc is plumbed by John Powelson in his chapter on "Oil Prices and the World Balance of Payments." He has compared the plight of the less-developed countries with those of the industrialized grouping; the former have been caught in a two-way vise, paying higher oil imports both for petroleum and for the manufactured goods produced in economically advanced nations where the higher oil prices are inflated in the process of value added.

Non-OPEC sources of petroleum have assumed greater allure for the major importing countries by those who see OPEC as a cartel that can be weakened by supplies from outside the Organization. Øystein Noreng has examined the complex and sometimes ambivalent nature of the relationship between non-OPEC petroleum exporters and OPEC.

A prominent issue that surfaces during the OPEC price-setting meetings is the erosion in the purchasing power of their oil-generated funds: To what extent should OPEC increases mirror inflation? Jawad Hashim, the Head of the Arab Monetary Fund; Massood Samii, an Economic Analyst with OPEC; and Mansoor Dailami, with the Energy Laboratory of Massachusetts Institute of Technology, provide three excellent chapters treating various aspects of this factor.

It is perhaps fitting that Herbert Hansen's clear-sighted call for international cooperation in energy concludes this volume. He believes OPEC should lead the way in this endeavor, as it can best defuse both short- and long-term tensions over nonrenewable energy resources.

In this editor's view, articles on OPEC tend to focus only on supply, demand, and price issues, with not enough emphasis on the linkages to development and related economic affairs for both importers and the OPEC states. The experience of the 1970s shows two facts: (a) the era of cheap energy of the 1950s and 1960s is over – at least for the next two to three decades until major technological breakthroughs occur, and (b) while it will take time to reverse the trend in supply and the price of energy now pegged to oil, there are avenues open to reduce the types of shocks caused in 1973-1974 and 1979-1980.

The lessons of the past, particularly those of the 1970s, will be enhanced by the future. The lesson of interdependence will be even greater, not only for the oil consumer but possibly even more so for the producer/exporter nations as industrialization and economic diversification efforts domestically are ever more closely tied to international trade, monetary policy, the transfer of technology, and the rational management of nonrenewable resources. For example, the massive industrialization programs in many OPEC states are based upon natural gas as a feedstock as well as an energy source, thus tying up a consistent crude oil output level as most often the natural gas is produced in conjunction with the oil (termed "associated gas"). Further, the petrochemical and refined products are likely to be heavily export-oriented.

Responsibility for the problem of the developing bloc cannot be assigned solely to either the advanced group of nations or to OPEC. More than 50 percent of some less-developed countries' foreign exchange goes for oil payments. Moreover, resource exploration in the developing nations is increasing with favorable results, since the energy resources largely have been untapped. Among the newer and non-OPEC oil producers are Sudan, Egypt, Mexico, Malaysia, China, Peru, and Ghana. A combination of OPEC financing, advanced-nation technology, and experienced labor can do much in locating and exploiting Third World energy resources to the benefit of all.

Another crucial lesson learned is that conservation is an effective means of reducing pressures on supply. Conservation, along with the general recession, saw oil consumption in the 21-member International Energy Agency grouping drop 7.5 percent in 1980 from the 1979 level of 35.5 million barrels per day. Because of ongoing conservation programs and the economic conditions prevailing worldwide, it is expected that consumption of oil will decrease another 1 to 2 percent in 1981.

Oil prices have been and are a factor in the inflation besetting all nations. While it may not be as great a factor as some people in importing countries assert, nonetheless, petroleum price hikes have a pervasive impact on any economy, especially the industrialized ones, due to the multiplier effect. Given the undeniable massive place of oil in world trade and the myriad tugs and pushes between and among the exporters and importers, energy management is here to stay as a major factor demanding global attention, concern, and solutions. Will we be more successful in a concerted effort to manage the nonrenewable energy source of oil in the 1980s than in the preceding decade? What kinds of measures can be instituted to bring equity to those nations depleting their primary source of income, while retaining a modicum of stability for the oil importers? Certainly a better understanding of the requirements and priorities of the OPEC nations will facilitate the structuring of mechanisms to balance the needs of both producers and consumers of petroleum. May this volume serve to widen the type of knowledge that is so basic to sound policy making on both the national and international levels.

OPEC: Twenty Years and Beyond

1

THE WORLD ENERGY OUTLOOK IN THE 1980s AND THE ROLE OF OPEC

*René G. Ortiz**

Introduction

History will record it was the Organization of the Petroleum Exporting Countries (OPEC) that first drew attention to the energy problem. However, it was not until 1973, when OPEC assumed the administration of oil pricing, that the world suddenly became aware of the "depletable and nonrenewable" nature of the oil resource. In proclaiming the exhaustibility of a commodity which, because of its cheapness and seeming abundance, had been taken for granted — and consumed accordingly — OPEC confronted the world in general, and the oil industry in particular, with an entirely new situation, one for which it was ill-prepared and, at first, unwilling to accept.

Needless to say, in the prevailing circumstances of the time the OPEC message was not well received. Several statements addressed to the consuming nations underlined certain unpalatable truths: that the oil reserves were the rightful property not of the international oil companies but of the producer nations; that the oil, which was being pumped at ever-increasing speed and which represented for many of the producers their sole marketable asset, would one day, in the not too distant future, reach exhaustion; and that, as a depletable resource, oil is and had been all along underpriced.

---

*His Excellency René G. Ortiz, Secretary General of the Organization of the Petroleum Exporting Countries (OPEC) headquartered in Vienna, earned undergraduate and graduate diplomas in engineering, systems analysis, and in economics and international law from the National Polytechnic School (Ecuador), Harvard Graduate School of Business, the Fletcher School of Law and Diplomacy (Tufts University), and the Institute of Superior National Studies in Quito. Since 1974 the author has headed various Ecuadorian delegations to OPEC meetings and price-setting conferences as well as serving in his country's representations to the VI and VIII OLADE (Latin American Energy Organization) sessions in 1977 and 1978.

History will also record that, following the initial unjustified violent reaction to OPEC's message on the part of the consumer nations and their governments, the voice of reason began to make itself heard in the councils of the major industrialized consumers, and the whole question of future energy supply became an issue of first importance. This new "energy-consciousness" led to a more respectful attitude towards OPEC and its policies. It came to be realized, if only reluctantly, that the producers were talking sense; that both producers and consumers must take measures to prepare against the day when there would be no more oil.

OPEC's case for reduced consumption; for accelerated exploration and exploitation of known and presumed reserves; for real efforts at conservation; and for a concerted drive to develop alternative sources of conventional and nonconventional energy in sufficient quantities to enable a gradual transition from oil to sources of a renewable nature – all these pleas increasingly began to fall on at least some receptive ears in the consumer countries.

The situtation today is but little improved on that which compelled OPEC's intervention in 1973. Admittedly, the price of oil has risen, but is still nowhere near its true value in real terms. The rate of oil consumption, while having decreased noticeably in the current decade with respect to the past, is still markedly higher than that of coal. At the same time the reserves have not increased as in the 1950s and 1960s but, on the contrary, have been marked in this decade by a sharp decrease with respect to the past. The figures in table 1 illustrate this issue.

Table 1

CONSUMPTION VERSUS OPEC RESERVES
(in percent)

	1950-1960	1960-1970	1970-1977
Rate of oil consumption increase	7.5	8.0	3.8
Rate of OPEC reserves increase	13.0	7.7	-1.1

Source: OPEC statistics.

This clearly indicates that exploration is not proceeding fast enough, while the exploitation of existing reserves of oil and gas, especially in OPEC member countries, is moving along quickly. Simultaneously, there is little evidence that genuine efforts at conservation are being made.

The purpose of this article is to examine the outlook for the coming years, with particular reference to the role of OPEC in its capacity as the traditional supplier of the bulk of the world's crude oil. The study will deal briefly with the history of oil's steady rise to become, in place of coal, the major source of energy, with its continuing predominance in the present day, and the emergence of serious interest in alternatives to oil. The article will also discuss the pricing mechanism of oil as a practical tool for the development and supply of other conventional and nonconventional sources of energy, and the need for governmental policies in consumer countries to encourage energy saving as well as legislation to implement conservation. The role of non-OPEC sources of oil and gas will also be touched upon, together with the delicate balance between supply and demand and the prospects of being able to achieve a smooth transition from oil to alternative sources of energy. Finally, proposals will be put forward whereby cooperation between OPEC and the consumers could lead to the world being able to avoid a potentially disastrous gap in that transitional period.

Oil versus Coal

Since the 1800s, the chief sources of the world's industrial energy have been fossil fuels, mainly coal, oil, and gas, all of which are of a nonrenewable nature. A look at the cumulative consumption figures clearly reveals that although coal has been in use for about 1,000 years, half of the quantities produced so far have been mined since the beginning of the Second World War, while half of the world's petroleum cumulative exploitation has occurred since 1956 only. Furthermore, the bulk of the world's consumption of energy from fossil fuels has taken place within the last 40 years. Until 1900, the contribution derived from oil and gas as compared with coal was negligible, amounting to a mere 8 percent as opposed to 89 percent. In 1950, oil and gas increased their share from 8 percent to 34 percent, whilst coal decreased from 89 percent to 59 percent. In 1960, the rates were almost equally balanced, oil and gas occupying 44 percent of the total energy supply and coal 49 percent.

Thereafter, however, the contribution of oil and gas steadily increased; in 1968, the figures were: oil and gas 57 percent and coal, continuing its downward trend, 36 percent. By 1972, the oil and gas share had reached 63 percent, while that of coal went down to 31 percent. In 1977, coal and oil and gas occupied 30 percent and 62 percent, respectively, of total energy supply. Furthermore, the rate of increase in oil and gas consumption was greater than the rate of growth in energy demand, while the reverse was true of coal.

Table 2 illustrates the behavior of the growth of total demand of primary energy versus that of oil and gas and coal in the period 1950 to 1977.

Table 2

GROWTH OF TOTAL PRIMARY ENERGY DEMAND VERSUS OIL AND GAS AND COAL
(in percent)

	1950-1960	1960-1970	1970-1977
Total demand	5.2	4.9	3.3
Oil and gas	8.10	8.3	3.55
Coal	3.5	1.0	2.00

Source: OPEC statistics.

A Cheap Resource

The reason why oil and gas resources came to occupy such a predominant place in the supply of energy is easily found: they were extremely cheap, easy of access, and unique in their scope of utility. It was hydrocarbons that provided the developed countries, through their command of the oil resources of the Middle East, with the cornerstone upon which they were able to build their industrial power. Indeed, for more than a quarter of a century after 1945, the nations of the West enjoyed a degree of uninterrupted economic expansion unparalleled in human history. One of the keys to the wealth thus generated was oil — oil in abundance and at ever lower prices.

From the producers' point of view, however, this situation was untenable and could not be permitted to continue. Thus OPEC came into being.

After 1973, the attention of the world was focused, as never before, on the problems related to natural resources and energy. Both the strategies of OPEC member countries and the solemn warning of the Club of Rome were instrumental in awakening the world to the startling concept that natural resources were limited, thus inducing governments, private institutions, and others to give serious thought to the availability and future supply of hydrocarbons.

Oil Is Still Supreme

The structure and consumption patterns of industrial society being what they are, however, and given the present price of oil as compared with alternatives, the resource continues to occupy pride of place among the sources of energy. *But for how much longer will oil rule the world?* With present world proven reserves of oil estimated at 650 billion barrels, and present growth rates of consumption amounting to 22 billion barrels per year, the oil resource will dry up within the next three decades. In the case of gas, the picture is hardly any brighter. At present growth rates of consumption, the reserves, amounting to 71 trillion cubic meters, will last for another 45 years.

The present proven oil reserves within OPEC member countries now stand at about 450 billion barrels. If the present OPEC production level (30 million barrels per day) were to be maintained, OPEC's output would start to decline at the end of this century, reaching exhaustion around the year 2025. Past and present estimates of ultimately recoverable reserves of oil and gas range extensively, but a reasonable consensus of opinion favors the figure of between 1,600 and 2,000 billion barrels, although some estimates, based on linear extrapolation of the past trend for the increase of such reserves, go as high as 4,000 billion barrels. However, further continued expansion on the same scale seems unlikely although there are good reasons for believing that, in time, expertise coupled with favorable economics will certainly allow additional resources to become recoverable.

With the estimated availability of additional conventional resources and the variety of unconventional sources of petroleum and gas, the age of petroleum could be considerably extended, enabling it to play a still important role in the transition to an energy economy based, hopefully, on renewable resources. It is with the conditions under which such a transition can be managed that today's energy debate is primarily concerned. In this connection, it is naturally important to know how long can, or must, the transitional period last, and for how long can we rely on oil and gas resources during this period. Obviously, the ultimate aim must be to augment the age of petroleum in order to permit a real breakthrough in the development of alternative sources of energy, especially the renewable ones, so that these can be made available to the extent necessary and at reasonable cost as the nonrenewable resources dwindle.

Conventional and Unconventional Resources

To reach the above objectives, it will be necessary to achieve: (a) an increase in the hydrocarbon resource base by conventional and/or unconventional means; and (b) rapid development of alternative sources of energy, especially those of a renewable nature. With regard to the first prerequisite, it should be stated that regardless of what can be considered as an accurate estimate of the ultimately recoverable amount of hydrocarbons, the extent and availability of oil and natural gas, the fuller use of conventional resources opened up by enhanced recovery, and the possibility of exploiting hitherto untapped unconventional resources — these are all governed by the efforts made and the success achieved in solving the various technical, financial and political constraints which cause a significant hindrance to the rapid development of additional reserves.

As far as the second prerequisite is concerned, due to the fact that prior to 1973 oil and gas prices were artificially maintained at unjustifiably low levels (which were just above the production costs in some areas), these exhaustible commodities — for some countries the only source of revenue — were undervalued for a considerable number of years and were over-consumed as if they were inexhaustible. Even following 1973-1974, when OPEC started to adjust its crude prices periodically, the previous pattern of consumption of the various fossil fuels continued to follow the same trend, whereby oil and gas were increasingly contributing to the world primary energy supply and coal behaving in the reverse fashion, as seen in table 3.

Table 3

OIL AND GAS CONTRIBUTION TO WORLD ENERGY SUPPLY
(in percent)

	1960	1970	1974	1977
Oil and gas	43.9	59.7	61.6	61.0
Solid fuels	49.0	33.9	31.2	31.1
Others	6.6	6.3	7.1	7.9

Source: OPEC statistics.

It is worth mentioning at this stage that OPEC's 1973 decision, in addition to deriving a more adequate value for its exports, was aimed at giving an economic incentive to the development of alternative sources of energy. However, such development has not yet gained the momentum envisaged, and it may well be that the present price of crude is still too low to offer the required stimulus for this undertaking at a faster pace.

For about a quarter of a century, nuclear energy has been regarded by many as the natural successor to petroleum and gas in providing clean and safe energy sources. However, the perspective of atomic reactor proliferation is now generating widespread and deep concern with regard to safety, while costs are increasingly being challenged. Furthermore, due to the limited utilization of solar energy and the fact that efforts and financial support are slowly being devoted to this field, no real breakthrough is anticipated in the near future. Nonetheless, it is hoped that those countries with financial and technological capabilities will direct more attention to this area, thus enabling the energy supply system to be based no longer on hydrocarbons alone, but on a multiplicity of energy sources, in particular those of a renewable nature.

Extending the transition period requires far more than development of new supplies or diversification of resources; it also calls for practical intervention from the demand side. The demand for natural petroleum is a complex issue because it is affected *inter alia* by prices, by the development of alternative sources of energy and by the changes in end-use technology which take place slowly. Conservation of hydrocarbon resources entails the capacity to use these resources in such a way as to avoid waste and their exhaustion before alternatives become available.

At this stage it is worth giving some thought to the various practices and measures adopted by major consumers towards the hydrocarbon conservation issue. OECD countries, while adopting sound, long-term measures towards their indigenous resources (such as the restraint on production in the United Kingdom, Norway, and Holland, the limited well production policy in the United States, or the exploitation of coal in both Europe and America), fail to apply the same attitude and sound concept towards foreign supplies of these exhaustible resources.

The conservation measures adopted so far do not constitute a real concession to the exhaustibility of these resources. Studies carried out on the energy habits of consumer countries show the possibility not only of drastically reducing the wastage of energy, but also, more generally, of organizing energy systems different from the present ones, systems capable of integrating a multiplicity of sources, both old and new, using the materials available to the fullest extent.

In view of this, major consuming developed countries should pursue a more vigorous policy aimed at: (a) eliminating all forms of waste stemming from irrational uses of hydrocarbons; (b) revising production and transformation structures in order to eliminate sizable existing losses in conversion and transportation; and (c) integrating into the system other energy sources that can contribute to the energy balance. A rational approach for consuming hydrocarbon resources entails a global integration of all parameters involved, based on a reasonable appreciation by the consumers of the requirements of the producers.

This may now be the appropriate moment to sketch out the role of the Organization of the Petroleum Exporting Countries (OPEC) in the future supply of oil. It is felt necessary that the life-span of OPEC's hydrocarbon reserves should be extended for as long as possible, so as to provide its member countries with the necessary time to transform their depletable deposits into permanent assets through development and industrialization and later to feed the industries created.

The necessity, however, to place future production in such a way as to ensure that OPEC can come to be regarded as a reliable, long-term supplier rather than a mere buffer on the international market, is becoming increasingly obvious.

It is worth mentioning here that after 1973, when people started talking about the drastic redistribution of resources to the advantage of our member countries, they omitted to mention the fact that these resources had not yet been transformed, except to a very small extent, into wealth. Today, some of our member countries with increased foreign exchange at their disposal are encountering serious difficulties in converting the value of their oil exports into real wealth, while other nations are becoming heavy borrowers in the international financial centers in order to be able to achieve their development aspirations.

Most OPEC states still require a long time to accomplish this transformation, while those that had a greater capacity for economic growth have seen this capacity diminish drastically due to the fact that unfavorable terms of trade have placed considerable constraints on their development. Surely everyone will agree that accumulating large bank deposits, quite apart from the possibility of using them to produce more income for the country concerned, corresponds neither to the best conservation criterion nor to the best development policy. In this regard, the developed nations have a moral obligation to assist in accelerating the development process of our developing countries through the adequate and timely transfer of modern technology.

In the period leading up to the exhaustion of the oil reserves, we need to have established not only self-sustained economic growth, but, independently of oil revenues, some of our member countries have to reach a stage where their nonpetroleum exports can pay for their eventual energy imports. Projecting the worldwide geographical distribution of coal, oil, gas, and other sources, it is expected that the energy source balance will soon shift from the noticeably rich areas in liquid and gaseous hydrocarbons (OPEC) to the present consuming countries, which are relatively rich in coal and have the greater possibility for developing alternative energy technologies. Therefore, in the long run, there is ample opportunity for a mutually beneficial exchange of commodities: OPEC continuing to export hydrocarbons for noble uses, some industrialized nations exporting in their turn to OPEC nonversatile forms of energy which are not indigenous to OPEC member countries. This complementarity can expand further in the respective industrial development of both parties concerned.

In this context, and given rational pricing of gaseous and liquid petroleum products, an orderly replacement should take effect, allowing petroleum products to contribute more to transportation and nonenergy uses of hydrocarbons. The best way to achieve this is by establishing a practical pricing of crude oil at levels not lower than the cost of providing a long-term supply of alternatives. It follows that at a later stage and to encourage the replacement of oil and gas as thermal fuels, these will have to be priced *above* the most costly energy alternatives. In this regard, it must be said that using a nonrenewable resource primarily as a fuel is an economic waste, and efforts therefore should be devoted to enhance its usage in the nonenergy sector.

Demand for OPEC Oil and Gas

It is only recently that consumers started to direct attention to the problem not of the pricing of hydrocarbons but of their availability. Forecasts pertaining to future energy demands, particularly that of oil, also have flourished recently. Of all these forecasts, energy figures based on reasonable assumptions gave the following breakdown for OPEC supplies until the turn of the century: 1980, 26-30 million barrels per day; 1985, 35-40 million barrels per day; 1990, 46-50 million barrels per day. From the reserves-cum-reservoir engineering points of view, the OPEC areas could physically sustain such levels of production, with some reservations as to the period after 1990 in this forecast. However, the question remains open as to whether the conditions will be met in the future so that OPEC could satisfy this demand. It is obvious that if OPEC were to satisfy these projected demands, the time-span till exhaustion of the reserves might not, in the longer run, be acceptable in view of the negative rate of growth of reserves mentioned previously.

A vital and influential factor which should not be disregarded is that of the growing internal consumption in OPEC states. At the moment this represents something like an average of 15-20 percent per annum, which indicates an ever-increasing rise in demand, thus depleting the availability of this commodity to the rest of the world. Furthermore, it is predicted that by 1990, internal consumption will have risen to such an extent that member countries' demand for crude oil will increase to some 6 million barrels per day, possibly tripling this amount by the year 2000, thus being roughly equivalent to the present consumption in the United States.

This means that for some member countries the day is not far off when their role will be transformed from that of exporters to nonexporters; indeed this will become a real possibliity by the turn of the century. In view of this, many OPEC states are confronted with the fact that the industrialized countries are awaiting a significant increase in the level of their exports — and this at a time when rapidly rising internal consumption is serving only to further reduce that level. The reason for this situation can be found in the fact that the consumer governments are either unwilling or unable to grasp the basic root of the problem which is simply that too little is being done to face the day when the only means available of properly allocating this depletable and nonrenewable commodity will be by using the price mechanism.

It is perhaps appropriate to mention at this point something about the utilization of natural gas. The gas reserves of OPEC member countries are in the order of 28 trillion cubic meters (174 billion barrels of oil equivalent), representing about 40 percent of total world gas reserves. Gas production in OPEC territories is in the order of 4 million barrels per day of oil equivalent, which represents 16 percent of total world gas production. It is worth noting that a considerable amount of OPEC gas is flared — 52 percent compared with 12 percent throughout the world. OPEC's flaring represents 71 percent of total world flaring of gas. In 1977 alone, for example, OPEC's gas production amounted to 268 billion cubic meters, more than half of which was flared; this represents a loss of about 2.6 million barrels per day, assets which will not be accounted for as part of their wealth.

It is only fair to observe here that the practice of flaring gas was initiated by the operating companies at a time when no consideration was given to its future utilization and attractiveness as a clean primary energy source. The discovery of gas along with oil has often been regarded as an inconvenience, and it was not rare to read reports of such discoveries being termed as unfortunate. In the same vein, it was not rare that some producing countries' legislations were penalizing utilization of gas by charging for the associated gas utilized, while flared gas remained unaccounted for, thus giving more incentive to flaring.

Companies equipped with the necessary technical know-how could work towards a greater efficiency in the utilization of such gas through cooperation with the appropriate authorities or national oil companies of OPEC member countries. Such cooperation, by increasing the hydrocarbon resource base, would give OPEC states a wider scope with which to meet the essential requirements of the consumer countries.

Due to the utter waste of flaring gas and the limited ability of OPEC member countries to consume it domestically, OPEC nations are faced with two possibilities: either to engage in a crash program for internal consumption of gas and exports, or gas reinjection projects, or to put an end to this wastage by linking production of oil to the associated gas flared so that optimization of gas utilization could be achieved.

The Role of Non-OPEC Oil and Gas

No assessment of the future energy outlook would be complete without reference to the contribution which non-OPEC areas can make to the total energy supply. Those areas are located primarily in: the U.S.S.R., China,

Alaska, North Sea (United Kingdom and Norway), Latin America (mainly Mexico), and South and Southeast Asia. In order to assess the actual and potential contribution of these areas, we shall now examine briefly the status of each.

U.S.S.R.: The Soviet Union today rates as the first country in oil production, and the second in gas production and proven oil reserves. Nevertheless, it has not yet reached maturity in the fields of oil exploration and production. Published data indicate that the Soviet Union produced 11.4 million barrels per day in 1978 and plans to reach a target of 12.4-12.8 million barrels per day in early 1980. In 1977, the Soviet Union exported about 3.2 million barrels per day of crude and products, mainly to the East bloc (about 29 percent of production), at the same time importing about 150,000 barrels per day of crude and products. Some estimates put the possible surplus that could be available for export to the CEMA at about 1 million barrels per day in 1980, assuming that production will then attain the rate of 12.6 million barrels per day.

There are also recent indications that the discovery of reserves is slowing down, and even taking into account new producing areas of oil and gas, the Soviets may be forced to reduce their future production rates. In short, all indications point to a huge potential of gas and oil reserves in the U.S.S.R., but there is no danger that the Communist bloc will become a burden on the reserves of OPEC member countries; neither does there appear to be any risk of mass replacement by Soviet oil of that produced by OPEC states.

China: In 1978, China produced 2 million barrels per day of oil. Three years earlier, it was predicted that China's production might reach 4 million barrels per day by 1980; later estimates, however, placed the figure at between 2.4-2.8 million barrels per day. If this level can be reached and oil production from the coastal fields of Taching, Shengli, and Takang, is developed, then it may be possible to export some 1 million barrels per day, 600,000-800,000 barrels of which would go to the Japanese market and the remainder to other countries, mainly in Asia and Africa.

The export of natural gas will depend, of course, on large investments required to finance the liquefaction plants. Most of the reserves of natural gas are located at Szechan in the interior and, at the moment, the gas is being flared due to lack of transport to industrial centers. In 1977, China produced 20 trillion cubic meters of gas and given the necessary technology, China might even be able to export gas to Japan. Nevertheless, it is not anticipated that China will play an important role in international oil supply before 1985, nor will it replace OPEC in this capacity; however, it will no doubt be able to keep up with the growing demand, especially of the region.

Alaska: Although Alaska has the largest known reserves of undeveloped natural gas potential (20 billion cubic feet) in North America, its transportation is presenting a stiff challenge due to the cold climate, remoteness of the area, and the permafrost. With American oil imports still growing, one can appreciate the interest of the United States in Alaskan oil. Supplies are now flowing through the trans-Alaskan pipeline which will eventually build up to 2 million barrels per day; it has been estimated that crude oil production from Alaska could reach approximately 3 million barrels per day by 1985. Of this quantity, 2.5 million barrels per day will depend on production from areas around Prudhoe Bay on the North Slope. Supplies from this source could account for 21 percent of total United States projected domestic production for 1985. If fully exploited by that date, North Alaska would experience a production decline of 0.6 million barrels per day by 1989.

The outlook therefore is far from being clear. It is doubtful whether Alaskan crude supply levels will ever exceed the 2 million barrels per day proposed capacity level. There is, however, the possibility of an exportable "surplus," because the oil will be going to the wrong place — the West coast of the United States — and currently there is no pipeline available to transport this "surplus" to the mid-West and Eastern regions of the United States where it is needed.

North Sea: North Sea oil is projected to contribute a fair amount to the oil requirements of Western Europe, its share in satisfying oil demand increasing from 1.1 percent in 1975, to 20.5 percent in 1980, 21.4 percent in 1985, then falling to 17 percent in 1990. Such a contribution without doubt will help considerably towards reducing the bill for imported oil and give some countries, such as the United Kingdom and Norway, oil self-sufficiency.

In the case of gas, the North Sea contribution to demand in Western Europe will rise from 19.6 percent in 1975, to 32 percent in 1980, thereafter decreasing to 30.2 percent in 1985, and 20.7 percent in 1990.

Western Europe satisfies most of its oil deficit from OPEC member countries, mainly those in the Middle East and North Africa. Its gas deficit is made up mainly by Algeria and the Libyan Arab Jamahiriya. The contemplated growth rate for gas demand surpasses North Sea production; therefore, the effect of North Sea oil on OPEC's crude oil exports is of more concern than gas. North Sea oil production is projected to increase from 1 million barrels per day in 1977, to twice that quantity in 1979, from then on steadily increasing to 3.2 million barrels per day in 1980 and 4.2 million barrels per day in 1985, after which it would decline to 3.5 million barrels per day in 1990 — all of which indicates an appreciable displacement of some OPEC crudes.

The OPEC crudes which will be affected to a greater degree are, therefore, the light African crudes and those from the Middle East with high distillate yields.

Latin America: Proven reserves of the non-OPEC Latin American countries at the end of 1978 stood at approximately 2 billion barrels of crude oil and 72 trillion cubic feet of gas. It is generally accepted that the bulk of crude oil and gas that will be available from non-OPEC Latin America countries will come from Mexico (oil, 73 percent; gas, 44 percent). In the remaining countries, the bulk of available oil and gas will be consumed internally and exports, if any, will be negligible.

Mexico, according to the most recent line adopted, intends to follow the path of a "cautious" producer. Accordingly, Mexico could choose to produce around 2.2 million barrels per day, thus being able to stretch production well into the next decades. Recent reports indicate that Mexico has not unraveled all its possibilities as far as oil and gas are concerned; certainly reports put out by PEMEX indicate that these possibilities are indeed enormous. However, from the OPEC point of view, it is envisaged that by the time Mexican discoveries arrive on the market, the oil market will be so tight that this contribution will be welcomed by OPEC countries as they will go some way towards relieving the difficulties that are predicted to arise in the next two decades, and most certainly in the period following.

South and Southeast Asia: Although the geologic potential for hydrocarbon reserves in the countries of South and Southeast Asia is high, the fact remains that only 10 percent of such reserves are, at the moment, proven. The remaining oil and gas accumulations have not been discovered and will not contribute substantially to production before 1985; after this date, all will depend upon policies adopted and investments made. Most authoritative forecasts regarding the level of future production from the area are estimated to be in the range of 1.5-1.7 million barrels per day by 1981, thus keeping this area as a net importer of oil.

Conclusions

The foregoing analysis has shown that, based on the various forecasts of energy and oil demands as well as future oil supply from the non-OPEC areas, it is believed that under normal conditions there should be no cause for concern until about 1985, as OPEC has a comfortable margin of oil export potential in addition to a wealth of gas reserves to reach that target. This means that the world can still enjoy a respite of less than a decade and, considering the lead-time necessary to either switch or adjust to new sources of energy, this time-span could be rather tight. What is called for now is a period of adjustment towards a more rational pricing of liquid and gaseous hydrocarbons with the aim of achieving that oft-mentioned "smooth transfer" to new patterns of energy consumption. This implies that consumers must accept the need for progressive increases in the price of oil in real terms, so that that price eventually attains its true replacement cost well in advance of the time when actual scarcity threatens the process of smooth transition to alternatives. Acceptance of this situation is in the interests of both producers and consumers: that which is inevitable must be faced, and the sooner the better.

In OPEC's view, the delicacy of security of hydrocarbons supply requires of the major consumers the adoption of a responsible and serious stand vis-à-vis OPEC's dual policy, namely, that of the legitimate protection of the OPEC revenue per exported barrel in the short term and the acceptance of the oil price mechanism as the most practical incentive for the development and supply of alternative sources of energy in the medium and long term. Thus, continuity of the world's development can be sustained.

In this process, increased efforts to seek out more and more areas of cooperation between producers and consumers should be made. Exchanges of manpower, for example, could promote not only an increase in the technological expertise of our peoples but also a deeper understanding among the peoples of the developed and developing countries of each others' problems. There should be increased trade between OPEC national oil companies and those of other countries. This is particularly relevant in the field of hydrocarbons, which encompasses a sphere of activities the national oil companies of the OPEC states have embarked on, but where they still lack a certain amount of technical know-how. A transfer of technology is therefore vital if member countries wish to take on the role of full-fledged producers of finished products. Surely such a transfer of technology is not too high a price to ask, in view of the sacrifices that OPEC nations continue to make in order to meet the requirements of the consumers. It goes without saying that this goal cannot be achieved without a change of attitude by the consumers towards the producers. It is indeed high time that the consumers view their relations with the producers with a substantial degree of open-mindedness so as to explore all avenues of cooperation that would lead to the optimization of the hydrocarbon resources. This attitude would imply that the industrialized countries accept without unjust constraints and hindrances the products from those OPEC refineries, petrochemical and other plants which must be built in the natural process of our industrialization.

Thus far, unfortunately, there has been little evidence of a genuine desire on the part of the consumer countries to establish a new and realistic relationship with the producers. The consumers could begin this process by overcoming at long last that ever-present hurdle which has thus far caused them to regard OPEC as an adversary. They should realize that, on the contrary, OPEC is, and must be, interested in the health of the industrialized economies in order to retain the markets for its oil and gas. Measures likely to engender distrust in the motives of OPEC should be avoided, since the history of the oil industry after this Organization assumed control over prices has clearly demonstrated that the exporter body's policies, while administrating its oil fully, comply with the moderation exercised of late with high consideration being given to the economies of the industrialized nations.

OPEC for its part is willing to play its role in the international commuity by becoming the instrument in breaching, not only the energy gap, but also the wide and growing gap between North and South nations, i.e., the developed and developing blocs. Once again, it is necessary to accept fully the fact that OPEC's policies are in complete compliance with the concept of interdependence, applied in its global sense, that is, a two-way traffic system of economic, social and development needs. Bearing this in mind, we would call upon the consumers to acknowledge that we, and they, are in the same economic boat, and that no effort should be spared to prevent that boat from sinking.

2

TWO CRISES COMPARED: OPEC PRICING IN 1973-1975 AND 1978-1980

*J. E. Hartshorn**

During the 1970s, two "price explosions" for OPEC crude oil occurred, both accompanied by political crises in the Middle East. In between those crises, OPEC governments managed or chose to make only a few, much smaller, increases in their unit revenues and prices. Since the increases did not even match current inflation from 1974 to 1978, the real price actually fell (though at the moment it is imposed, any price increase seems real enough). For the 1980s, should one expect the same to hold true? Outside periods of crisis, will OPEC again have difficulty in raising real prices gradually but regularly, as its long-term strategists feel it now should do? Even if the answer were yes, it would say nothing about the likelihood of further abrupt increases, because further political crises in the Middle East must be reckoned as highly probable in the '80s. But a positive answer *would* express a view about how much market power OPEC still has or, at any rate, how much power it seems likely to exert. So a brief comparison of OPEC's pricing in and immediately after these two political crises may be instructive.

At the time this paper was completed in June 1980, it seemed possible, if not likely, that Iran's turbulent policy would be upset again in conflict with the United States or Iraq. However, there was not much more that sanctions or even a blockade could do to Iranian oil exports by then. The exports were already down to perhaps less than 1 million barrels per day (b/d), compared with 5 million b/d in mid-1978. The surplus of supply potentially available elsewhere over effective demand in the crude oil market of spring 1980 was probably two or three times as large as the remaining Iranian exports that

---

* Jack Hartshorn, a Vice President of Jensen Associates, Inc., Boston, has had a long career writing about and consulting on international oil economics and politics. The author of a number of studies, his most recent volume is *Objectives of the Petroleum Exporting Countries* (Nicosia, Cyprus: Middle East Petroleum and Economic Publications, 1978). Mr. Hartshorn also has been a Visiting Professor at the School of Oriental and Asian Studies, University of London, England.

might be cut off from the market. So the effects of a renewed crisis there upon the market might be considered very limited in strictly economic terms. Politically, however, a renewed Iranian crisis was still liable to affect the market by contributing yet again to all buyers' anxiety and some sellers' self-confidence.

Profiles of Crisis

In the amplitude of their effects upon OPEC and world crude production, the two crises were notably similar. Indeed, it is difficult to separate them on a single graph because the fluctuations in supply that they wrought happened at roughly the same level of OPEC production. Even the timing was only a couple of months out of phase. The decline in OPEC output from November 1978 to January 1979 – from a strike in the Iranian oilfields to a shutdown as the revolution erupted – was never quite as sharp or as deep as from September to November 1973, during the Arab oil embargo. (Production in autumn 1978 had never been quite as high as in autumn 1973, however.) In both periods, it took about six months for OPEC production to regain its pre-crisis level, but in 1979, *world* production took only about three months to recover. There was more non-OPEC supply available to fill in the gap in production. A comparison such as that in figure 1 is a reminder that while OPEC output in 1978-1980 was only at the same general level as in 1973-1975, world output had increased. Since that earlier crisis, there has been slow but substantial growth in world oil production: in net terms, all of it has been outside OPEC.

The political crisis affecting Middle East crude supplies dragged on much longer in 1979 than in 1973. As noted above, the pressure on supplies did not. In both cases the initial upsurge of prices reflected the understandable fear of importers that the crisis might reduce world crude supply sharply for a significant period, but in neither case did that happen. Both crises are almost lost in annual statistics. Following the Arab embargo, OPEC production in 1974 averaged less than 1 percent lower than in 1973, and world production of crude was virtually unchanged (admittedly, in the few years before, OPEC exports, particularly of Arab crude, had been rising rapidly). In 1979, even OPEC production turned out to be higher than in 1978, in spite of Iran's brief shutdown and prolonged reduction of exports, and world production reached a record 62.8 million b/d. But the expectations – and, perhaps even more, the fears of political uncertainty – were what mattered.

The 1978-1979 political crisis was not as unexpected as the October War and the oil embargo of 1973 had been. The proportionate increase in OPEC's current prices was not as large; prices "only" doubled, as against quadrupling. Real prices rose by about 50 to 60 percent, and, as figure 2 shows, it

Figure 1

WORLD AND OPEC CRUDE PRODUCTION DURING CRISES
(by month, beginning in July)

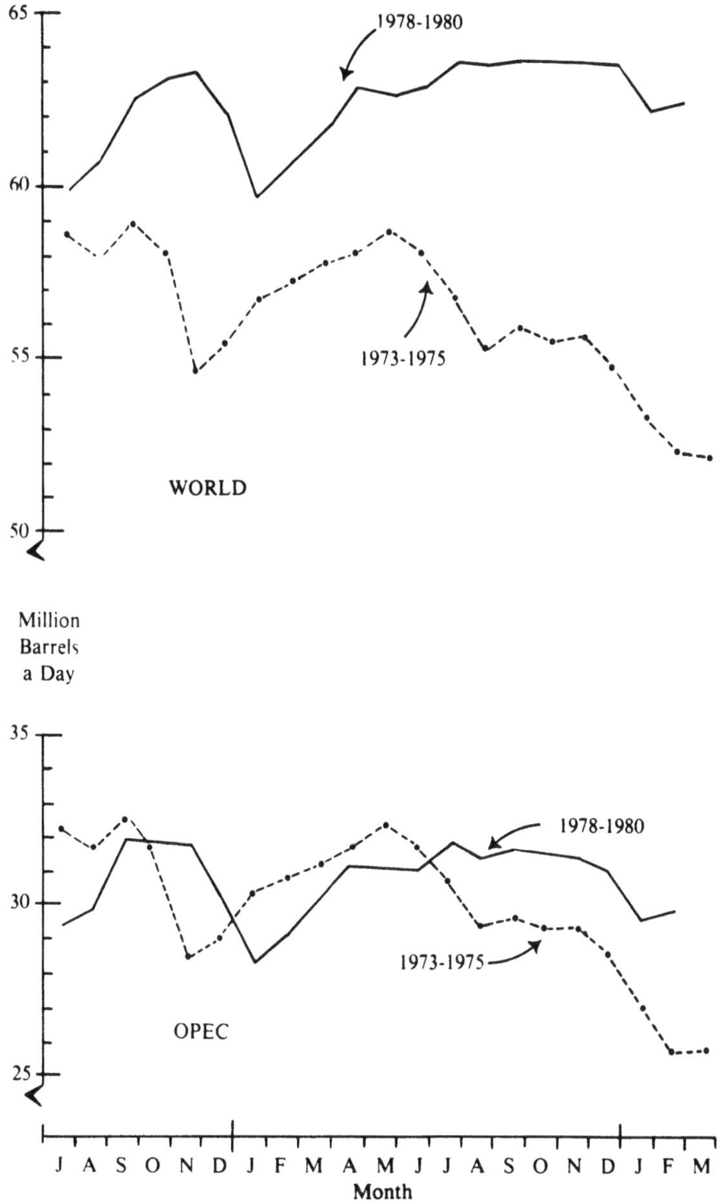

Figure 2

REAL PRICE OF OPEC CRUDE, 1978-1980
(Arabian Light, f.o.b. Ras Tanura)

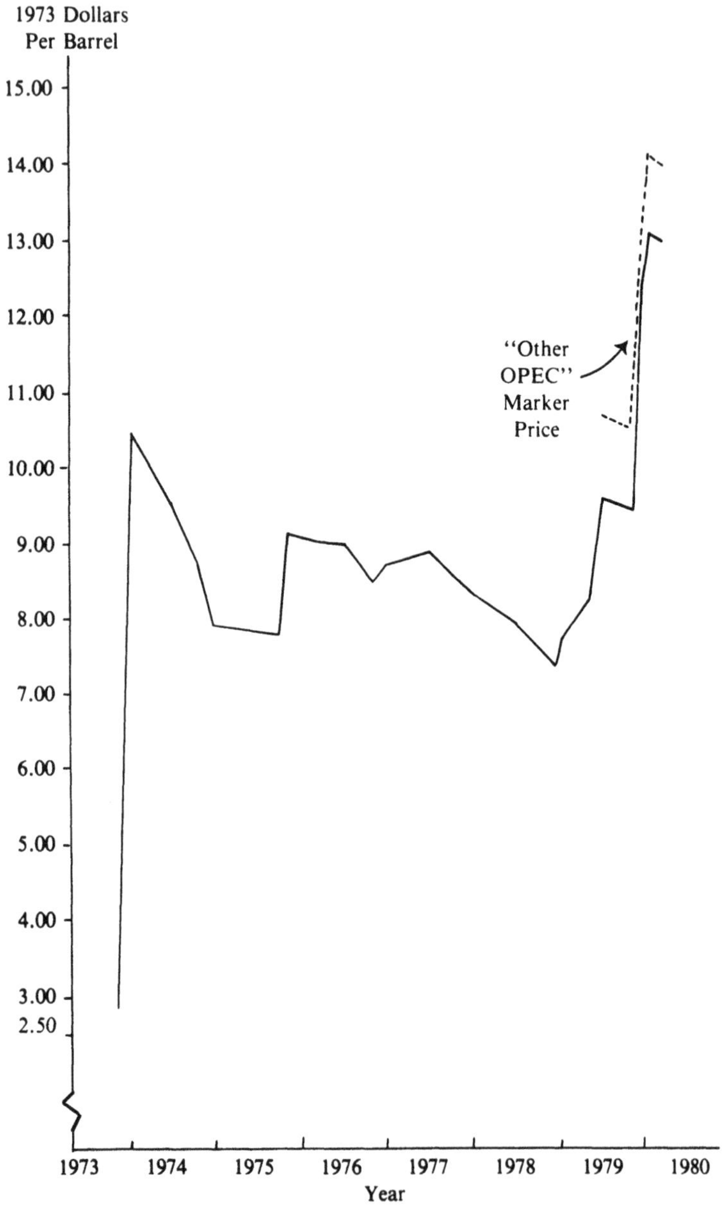

was not until the end of 1979 that they exceeded the levels reached at the beginning of 1974. But the absolute increase in current dollars per barrel was much larger; it brought back the world balance-of-payments problem of OPEC surpluses, which against all expectations had dwindled to almost nothing by 1978.[1] It speeded up a radical, irreversible shift in commercial channels of trade for crude oil to those directly involving OPEC governments, across a wider front than ever before, in the world's biggest export-import businesses. It also introduced a new element into the pricing of OPEC oil: the cost of insecurity.

Neither of the price explosions was initiated by OPEC. In each period, crude prices in the open market were already exceeding those charged by OPEC (or those determining its member governments' revenues per barrel)[2] before OPEC first reacted in October 1973 and again in December 1978. Then came the political crises (unexpected in 1973 but already widely anticipated in late 1978) and the temporary cuts in total OPEC exports. The crises and cuts sent open-market prices soaring even further. Then OPEC raised its own prices again – repeatedly in 1979, though it never managed to match prices in the spot market. In neither crisis year, however, were these further increases explained by OPEC spokesmen simply in terms of catching up with open-market or spot prices. In December 1973, the OPEC spokesman (who at that time happened to be the Shah) justified a further increase in terms of that recurrent but elusive concept "the cost of alternative

---

1 OPEC economies' imports – in fashionable jargon, OPEC's absorptive power – rose faster than anybody expected in the period 1974-1978; since oil revenues declined in real terms, the surpluses were steadily reduced. In 1979, revenues soared, and OPEC imports – again unexpectedly – seem to have been no higher than in 1978.

2 In 1973, it was the posted (tax reference) prices for Persian Gulf crudes that were raised, unilaterally for the first time, and very sharply, by an OPEC ministerial committee. Indeed, the raise was 70 percent. But it was designed to restore a traditional relationship between the posted prices, and hence the royalties and tax revenues based on them, and prices actually realized from sales to third parties. Those prices had traditionally been about 40 percent lower than posted prices, but in the summer and autumn of 1973 had actually risen to exceed them, thus reducing the governments' shares of the total profits made (or imputed) from selling crude. That OPEC move was thus designed to restore a traditional formula between open market prices and the revenues per barrel that OPEC governments got, in response to a strongly rising market.

The OPEC price increase in December 1978 may have had little to do directly with the rise in open market prices that had occurred in November because of a strike affecting Iranian exports. It was negotiated between OPEC "hawks" and "doves" as a compromise, a gradual quarter-by-quarter increase designed to do little more than offset the continuing erosion of the real value of OPEC's current government selling prices – in the knowledge that open market prices were higher – but without directly seeking to match them.

energies." In mid-1979 there was indeed talk of the very reverse of chasing spot prices: agreement on a price ceiling to stop official selling prices from being drawn upward by (and in turn further raising) the spot market. Nobody outside OPEC expected it to work, and it did not. But few OPEC governments disregarded the ceiling immediately or were publicly opposed to the principle. Indeed, the concept surfaced again in Algeria in June 1980.

Price in the Aftermath

In both crisis periods, world supplies quickly became ample again. Product prices declined, as did the "netback" refining values of crudes – first to below the spot prices for crude, and later to below the official selling prices at which crudes were obtainable from OPEC governments. Then the spot prices for crude weakened. By mid-1974, a number of OPEC governments had accepted actual reductions in some elements of their official pricing, for example, in the "buy-back" prices at which former concessionnaire companies had purchased some of their "participation crude" for resale. At that time, only a few OPEC governments were in fact selling much crude directly. The "offtaker" companies' average crude costs, which derived from tax-paid costs on the "equity crude" and higher prices on the participation crude that they resold, were more important than the posted prices formally announced by most OPEC governments. Then, as in 1980, OPEC was considering reunifying prices; on paper at least, there were even more disparate elements needing to be brought together. By mid-1974, therefore, there was some government-sanctioned weakening of official OPEC prices. At this writing, mid-1980, there has been no sign of that happening in the aftermath of the 1979 crisis; some governments continue to raise prices.

The record of prices, however, cannot be considered by itself in assessing the effect of downward market pressures on OPEC crude in 1974. Throughout most of that year, in spite of the erosion of prices, OPEC conferences continued to impose increases in royalties, tax rates, and other elements of "government take" and to seek larger shares of government participation. In November 1974, Saudi Arabia, Abu Dhabi, and Qatar unilaterally made further sharp increases in royalty and tax rates, and at a December conference, the rest of OPEC agreed to make comparable increases through their various patterns of pricing and unit revenues. (Conformity of formulas for unit oil revenues was never easy to arrange among 13 countries whose ownership of their industries varied from nil to 100 percent, even outside the periods of dispute and two-tier, or nowadays multi-tier, pricing.) The royalty and tax increase policy, forestalling other OPEC proposals, actually included a *reduction* in the formally posted price and suggested that the country's national oil company, Petromin, might be prepared to sell crude to third parties more cheaply than it sold incremental volumes to its Aramco offtaker companies. The fact remained, however, that on most of its export volume, the net effect of Saudi adjustment in both directions (royalties and taxes up, formal prices down) represented another big increase in government revenues per barrel. By the end of 1974, in spite of the weakness of the market, the Saudi government was getting $10 a barrel, compared with $7 at the beginning of the year.

That was the general pattern in 1974, even though during seven or eight months of the year, inflated product prices had been collapsing and market pressures had shown some significant effects upon certain published prices for crude. Government revenues per barrel went up by another 40 to 50 percent – little indeed compared with the crisis increases in late 1973, but increases *against* market pressure nevertheless.

OPEC governments were asserting their dominance over international crude oil pricing during that period, yet the dominance was mostly exerted indirectly. Although the OPEC governments acted in part directly upon price levels, to a large extent they acted upon the terms by which foreign companies acquired and sold the vast majority of the oil (it was sold partly as crude to third parties, but mostly as final products through their downstream networks). Direct commercial contacts of the governments and their national companies with oil customers were limited to a relatively small share of the business. During 1974, OPEC governments were still operating largely by squeezing economic "rent" out of the profits made by their former concessionnaires, who were then in the process of becoming company contractors. That had been the traditional mode of OPEC pressure up to the early 1970s – extracting more of the "producer's rent" and leaving the companies to recoup as much of this as they could by raising final prices. Only after 1973-1974 did the OPEC governments begin to shift their attention to the new goal of getting more of the consumers' and importing governments' surplus embodied in the final prices for oil products.[3] During 1974, OPEC's pressures on prices and on revenues per barrel (alternatively or in combination) displayed a mixture of the two approaches. Since then, there have been more signs of this mixed approach, including trends in 1979 and 1980.

In the recession of 1974-1975, some governments that had overpriced their crudes in relation to the basic OPEC marker crude price and whose company offtakers had cut their export volumes were obliged to accept price reduc-

---

3 Professors James M. Griffin and Henry B. Steele, in a penetrating if schematic analysis of this change in OPEC government pressures, put the change earlier. In practice, it seems to me that the two forms of pressure from OPEC continued to overlap for quite a time (see their *Energy Economics and Policy,* New York: Academic Press, 1980).

tions. The governments of the Middle East and Africa, producing some of the higher priced light and sweet crudes, as well as Kuwait, producing heavier and higher-sulfur crude, thus shared much of the 1975 downswing in demand upon OPEC with Saudi Arabia, producer of the marker crude, Arabian Light.[4] Saudi Arabia played less of a "swing supply" role during this period of weak demand than many outside observers had predicted. This was because of the difficulties that OPEC had encountered (and still faces) in assessing the "right" range of price differentials for the 30 to 40 different qualities of crude that its members sell in substantial quantities. But the differentials from the OPEC marker price succumbed to market pressure, not to the marker price itself. During much of 1975, to minimize yields of heavy fuel oil (for which demand had particularly weakened in the severe industrial recession in most OECD economies) there was a shift in demand toward the light, high API° gravity African crudes. By summer 1975, demand for crude strengthened slightly; however, this demand seemed to be for short-term stockpiling in anticipation of another OPEC price increase. And that increase came in due time: Saudi Arabia raised the price of its marker crude 10 percent, and other producers, seeking to fine-tune their differentials, raised their prices by various parallel percentages.

OPEC Pricing and Market Expectations

The increases in OPEC government takes in 1974 and in OPEC prices in 1975, during a fairly sustained softening of demand, seem to have demonstrated some power, simply on the basis of government say-so, to move contrary to market pressures. It is difficult to explain those increases as obvious responses to the balance of demand and supply in the market. In mid-1975, certainly, the market reflected expectations that OPEC would be able to impose an increase in spite of weak demand. But buyers' expectations would have been primarily concerned with the use of OPEC power – market power or simply political power – rather than with the strength of market demand and the supply available. The nature of buyers' expectations has changed. Today, guesses about future OPEC government behavior seem indispensable for all purchasers attempting to judge the future course of prices, quite apart from their ordinary judgment of future market demand.

---

[4] The marker crude is a 34° API Middle East crude, in principle Arabian Light. But during periods when Saudi Arabia has been at odds with other governments in OPEC, as at times in 1974 (on royalty and tax rates) and in 1977 and 1979-1980 (on prices), the other member governments have used a marker crude price for their benchmark, which has not corresponded with the price at which Saudi Arabia has sold the actual Arabian Light.

The two, obviously, interact. But OPEC behavior seems difficult to predict from market expectations alone.

In 1979 there was a much longer lag than in 1974 between correction of the widely expected – and exaggerated – commercial shortage and any response on the part of crude prices. Indeed, although stocks reached record levels in the winter of 1979-1980 (which turned out to be a mild season), and declining product prices weakened the refining values of most crudes before the turn of the year, OPEC governments continued to raise their official selling prices well into the summer of 1980, even while spot prices for crude were simultaneously coming down.

In late 1979, distortions within the official price structure had probably helped to widen the spread between official selling prices for crude and spot prices, along with the variety of prices in between (official prices plus temporary surcharges, exploration premiums, fees for contract supplies, etc.). Most OPEC governments were selling sizable, albeit shrinking, proportions of their crude exports to major company offtakers at official prices, often with discounts or other preferential elements in the contracts. When Saudi Arabia and one or two other member governments did not raise their official prices as high as the rest of OPEC wanted, their particular offtakers enjoyed at least temporary extra "subsidies" per barrel (though some of these were later cancelled retrospectively). Most of the major companies concerned, however, were having their export volumes reduced, eventually to less than what they needed for their own integrated downstream refining and marketing systems. To make up incremental requirements here and there in their worldwide systems, these offtakers could readily afford to pay whatever was demanded for spot crude supplies (most of their competitors were paying those spot prices for much larger proportions of their crude supplies). The restraint that led to two-tier pricing, as has been observed, probably did moderate the average level of OPEC's official prices in 1978-1980, and hence the level toward which spot prices for crude began to decline early in 1980. But it also had some complex effects upon the pattern of OPEC supply.

During most of 1979, Saudi crudes were cheaper than other comparable OPEC crudes. Therefore, they naturally appeared to be sold out most of the time.[5] This seemed to apply to the heavier, and less desirable, medium and heavy Arabian crudes (which were relatively more expensive), as well as to Arabian Light itself. Anyone lucky enough to get Arabian Light at from $2 to $6 a barrel cheaper than other average OPEC crudes had no reason to

---

[5] During much of 1979, no actual figures for Saudi production were published regularly. The trade press simply had to publish estimated figures on the assumption, which was probably fairly safe, that the country "must have been" producing at its maximum allowable level.

cavil at taking the proportions of heavier crudes that purchasers were required to lift along with it. During much of 1979, therefore, if there was any "swing supply" in OPEC, it may have been Kuwaiti or Iranian crudes (the Iranian exports were, at times, however, constrained by temporary interruptions). Demand slackened earliest for these heavier crudes because of shifts in the pattern of product demand. The U.S. government, believing it was faced with a temporary "natural gas bubble," sought to persuade utilities to use gas instead of heavy fuel oil. Because demand for electricity was slackening anyway, residual fuel oil seemed to become briefly unmarketable in the United States. Thus U.S. import demand[6] concentrated more than ever on the light African crudes, and the decline in general demand consequently affected the heavy crude suppliers of the Persian Gulf (as well as the Venezuelan source of crude and fuel oil) even more.

Another element that probably encouraged some OPEC governments to keep raising their official prices, even after spot prices for crude had begun to decline, was the profound shift between channels of trade in international oil. From offtaking perhaps 75 percent of OPEC crude exports in 1976 and supplying the majority of third-party buyers as well as their own integrated systems, the major international companies dropped to somewhat less than self-sufficiency by early 1980. They were no longer fully able to meet their own downstream requirements with crude under secure contracts. This left them competing for supplies with other party customers, rather than supplying them. In effect, the third-party market is now supplied almost wholly by OPEC national oil companies. And a host of smaller customers who had been accustomed to getting their crude through dependable and easily negotiated contracts with the international majors suddenly have had to begin buying directly from these national oil companies.

Insecurity of Supply

The new sellers (OPEC national oil companies) took over in a situation of apparent shortage and great bargaining advantage. When they sold crude at official prices, they were indeed doing most customers a favor. A number of these customers – notably importing national companies, some from poor nations and some from rich ones – in return could offer favors of one kind or another that the OPEC governments considered worth their while. New private customers had nothing comparable to offer, and national companies, aware that their crudes were fetching considerably higher than official prices

---

[6] In spite of genuine economies in the use of oil in 1979 and a reduction in total oil imports, imports of crude did rise slightly.

on the open market, did not hesitate to impose surcharges, to reduce volumes, and to seek other inducements. The circumstance in which customers began to buy from these new sellers was the reverse of that famous "security of supply" (intangible, perhaps sometimes overrated, but very real in the minds of private company purchasers and importing governments) that the majors had formerly offered. Buyers' behavior in 1979-1980 seemed to display just the reverse: a growing insecurity about getting regular supplies.

This insecurity may derive in part from the initial shock of 1973, which was a reaction not only to the size of the price increases that OPEC suddenly imposed, but also to the fact that OPEC governments should be able to set prices at all. The effects of those realizations upon business confidence throughout the industrialized oil-importing countries may have been as significant for the world economy as some of the other, allegedly more measurable, consequences of that OPEC action. However, the embargo came to an end fairly quickly and was perceived not to have cut supplies as much as had been feared (particularly in the United States, its main object). Supply, fully restored, still came mostly through the familiar channels of the major companies. The higher prices were passed along. Even the OPEC financial surpluses did not turn out to be as menacing as had been predicted.

But from the crisis of 1973-1974 and the years of seemingly irresistible bargaining power that OPEC governments have shown, there remains an intimidating residue of uncertainty that worries oil importers. This has not been allayed in any sense by the wave of anxious projections of long-term energy demand and supply that has taken up so much of the time of energy economists. In spite of increasing elaboration, the conclusions of those projections have seemed highly susceptible to fashion, swinging repeatedly between optimism and pessimism. At the time that the Iranian revolution upset supply, most oil importers were probably influenced by strongly pessimistic projections, prophesying an energy crunch in the mid-1980s (though in fact, predictably, the fashion had already begun swinging the other way). However, one thing that all such projections, whatever their tenor, had to emphasize was the high degree of uncertainty affecting all expectations – long, medium, or short – about the oil market.[7] On balance, they have probably tended to widen the general perception of the uncertainty inherent

---

[7]See J. E. Hartshorn, "Energy Expectations and Uncertainties," *Middle East Economic Survey* (Nicosia, Cyprus: Middle East Petroleum and Economic Publications), Supplement, November 14, 1977, pp. 1-7.

in future oil supply, not in any sense to reduce it.

In 1979-1980, when the international majors retreated from their intermediary role as sellers in the third-party market to concentrate on their own needs, a further element of uncertainty for other buyers was introduced. Pricing in early 1980 showed that buyers (including some majors particularly bereft of their former preferential supplies) were prepared to pay quite a significant price for any apparent reduction in the uncertainty of their supplies. For example, in order to obtain term contracts (seldom for more than 12 months), major and other companies were prepared to pay not only government selling prices for crude that spot prices would not cover at the time, but various surcharges as well. Smaller companies were vying with one another to offer nonprice incentives to national companies in order to obtain longer-term guarantees of crude supply – incentives such as joint ventures in refining and other industries in which OPEC governments might be interested, both in the OPEC countries and in importing nations.

The contracts for which these buyers have been prepared to pay premiums, representing what might be called a "cost of insecurity," are seldom long-term. They usually stipulate specific freedom for the supplier to adjust prices unilaterally and, at least implicitly, both sides know the supplier is free to provide less crude than the volumes specified in the contract if some other disposal suits him better. From their experience in 1979, many purchasers may be skeptical about the actual degree of security inherent in such contracts with national companies. But obviously they value it as a great deal better than nothing. Purchasers were prepared to pay those premiums during a period when stocks downstream were at record levels, the interest costs of holding them were uncommonly high, and spot prices for crude had begun to decline because of falling refining values reflecting a weak market.

During some of the years between the two oil crises, notably during 1976-1978, the bargaining stance of buyers often seemed rather intimidated when dealing with OPEC governments and their national companies. Most of the majors, enjoying privileged access to crude, were beholden to at least some of the governments they did business with. Smaller companies, not accustomed to bargaining across the table with sovereign governments, seemed almost overawed. Oil-importing governments, while also enjoying sovereignty, had nothing to offer that they could deliver that OPEC governments wanted as much as the importers wanted oil. During 1979-1980 that advantage of national sovereignty may have been concentrated into one element of bargaining. The national companies now handling a far larger share of the business – in exporting countries outside OPEC as well as inside – can, if they choose, offer medium- to long-term guarantees of supply; nobody else can.

Once-For-All Changes: Nearly Complete?

Following these political crises, both of which raised prices abruptly, OPEC governments were able to raise their basic prices (or revenues per barrel, which drove prices up) for a time *against* the apparent statistical position of demand and supply in the market. After 1973-1974 this practice was facilitated by the fact that they could exert further pressure to alter the terms upon which their concessionnaire-contractor companies were allowed to retain technical management of the most important supply sources. During 1979-1980, a change in the structure of the trade, shifting most third-party sales into the hands of national companies, made buyers ready to to pay high costs, regardless of the current state of the market, for any assurances of longer-term supplies from new and seemingly quite unpredictable sellers.

Both of those changes are once-for-all, and they are nearly complete. Elements of private ownership remain in several OPEC nations that have not decided to nationalize fully. The most important remaining bastion of private company presence, Aramco, is in a country that in principle made the nationalization decision in 1974. Saudi Arabia has never proceeded to the formal takeover, but since late 1976 it is said to have been applying the financial provisions of a draft agreement. It is not easy to see how the terms of any final Aramco takeover could enhance Saudi government power in pricing its oil. The same is true of the other countries that still employ companies as contractors of one kind or another (which have, in any case, less power in general OPEC pricing). The other change – the shift of trade to national company channels – has not far to go either. Completion of both changes may involve the final loss of preferential terms of access to crude for former concessionnaires, who might thereafter have to get crude from those nations on the same terms as other buyers. That will significantly affect the fortunes of certain of the major companies concerned. But again, it will not necessarily alter OPEC's power to maintain and increase prices, unless buyers feel that complete dependence upon national companies will further increase their insecurity of supply.

Before the Iranian revolution, many analysts of world oil, including more than one OPEC oil minister, were expecting the market to weaken from mid-1978 to about 1981. Certain non-OPEC supplies were due to increase quite rapidly, and have done so. A cyclical downturn in the world economy, with most of the industrialized economies moving into recession, was expected about 1980, and since 1979 economists have been predicting that it will be more severe and perhaps more prolonged than they first thought. Whether or not the world economy can begin expanding again in 1981-1982, most current projections suggest a slower growth rate out to 1985 than was hoped for earlier. OPEC spokesmen continue to lecture the OECD countries about their duty to save energy, and the OECD governments, through the International Energy Agency, continue to upbraid themselves for not saving enough. The fact remains that since 1973, allowing for recession, energy usage per unit of gross domestic product (GDP) in those economies has been reduced more than most observers would have guessed. If these energy savings continue to be achieved during a period of what looks like painfully slow economic growth until the mid-1980s, then "net demand on OPEC" – the bottom line of everybody's energy projection – might hardly increase between now and the mid-1980s, perhaps even throughout the decade.

Strategy Needs Management

In order to achieve its long-term strategy committee's definition of a floor price for OPEC crude (which is quite a modest price in real terms), the Organization might have to achieve increases in current prices of perhaps 15 percent per year.[8] Such a rate of increase is not intended to be obtained by infrequent sharp spurts in price; the committee would prefer predictable regular quarterly increases that everybody could get used to. The element of predictability in price might be welcome to many customers of OPEC crude, provided they could also achieve some degree of predictability of full supply. It is possible, however, that in the weak market considered above (*if* it develops) one of the only ways OPEC could gain such predictable price increases might be, specifically, by keeping supply rather unpredictable. Predictability for both supply and price might hardly be compatible, if the price needs the support of buyers' insecurity.

That sort of trade-off implies more management of OPEC's level of supply than ever before. Nobody is now seriously suggesting collective "production programming"; few governments would even consider it. The OPEC strategy committee assumed that the Organization's goal could be achieved through each government's conduct of its own production policy in its own national interest. That kind of "conscious parallelism" would certainly seem compatible in principle. The enlightened self-interest of most OPEC governments at least would generally work to the collective (special) advantage of

---

[8] The committee let it be known in February 1980 that it would recommend that OPEC set a floor price moving in line with three indexes: (a) an adjustment for inflation based two-thirds on the IMF index of exports of manufactures and one-third on the consumer price indexes of OECD countries; (b) an adjustment for changes in exchange value related to the former "Geneva I" index used to adjust OPEC posted prices in 1972–1973; and (c) an increase in real prices (thus adjusted) related to the average growth in GNP in the industrialized countries of OECD.

all of them. But it would be easier at some times than at others. Holding allowables unchanged while demand is rising, so that prices rise, is one thing; sticking to official prices when demand softens and accepting one's share of the loss in sales is another.

OPEC governments have had some hit-and-miss experience with both market situations and in general have been fairly successful. But to achieve steady increases in real price while demand is not rising will presumably require some actual and repeated downward adjustment of allowables and, hence, of the total OPEC supply being offered. That may not be impracticable or indeed illogical (consider the "backward-sloping supply curve" that some of its critics allege already represents OPEC's economic behavior), but it may require rather more consultation to achieve parallel downward adjustments of allowables, unless only one producer (guess who) were to be expected to assume this whole responsibility for keeping the market steadily rather tight. Not many OPEC governments other than Saudi Arabia may be willing or able to squeeze their production allowables more than they have already done. At some time in the early 1980s, moreover, one must hope, for everyone's sake (including Iran's), that Iran will again be exporting through one channel or another 2 or 3 million barrels of crude per day. Furthermore, for OPEC to be seen as constantly tightening the screws to *reduce* supply in order to raise prices would present a public relations image rather different from its current one of not being willing to waste its countries' natural resources simply to satisfy the ever-increasing appetites of the profligate industrialized West. (In the circumstances being postulated, those economies would be wasting less energy and possibly not becoming as rich.) All of OPEC is sensitive to public relations. Saudi Arabia, moreover, is deeply concerned about the future progress of the world economy and its own political relations with the United States. So the practicable method of achieving the predictable increase in real oil prices that OPEC's long-term strategy committee proposes (in market circumstances for the early 1980s that are, let us say, plausible) might thus put a highly unpopular and uncharacteristic onus of production policy upon the very government that sponsored the committee.

The market circumstances I've described are plausible but very far from certain. It would be pleasantly surprising if the world economy were to achieve anything better than a very low rate of real growth between now and 1985. Unfortunately, it would be less surprising if energy use per unit of GDP in the main oil-importing economies ceased to decline as much as it has, somewhat mysteriously, since 1973. Instead, energy demand may rise rather more, even if the world economy continues in low gear. Hence, net demand on OPEC would recover earlier; allowables might not need to be screwed down to make a floor price slope upward; OPEC might get what its strategists want and the OECD might get nothing worse than what its governments may privately consider fairly acceptable.

Conclusions

This chapter has attempted to assess what is practicable, not what is likely. It has considered the question of whether, in the absence of further political crises, OPEC will again have difficulty in raising real prices in the early 1980s, as it did in the mid- to late-1970s. The answer seems to be twofold: OPEC has the market power to achieve such an increase, provided that Saudi Arabia is prepared to lead other member governments in managing its own production allowables; and since Saudi Arabia has sponsored a gradual increase as an OPEC objective for the 1980s, presumably it must be prepared to manage output accordingly. However, the condition assumed for the question – that further Middle East political crises will be absent during the early 1980s – seems at present too much to hope for. Thus, this question about OPEC's crisis-free market power may never be answered in practice.

3

SOME LONG-TERM PROBLEMS IN OPEC OIL PRICING

*Robert S. Pindyck**

Introduction

In several recent articles I have examined OPEC's oil pricing problem on the basis of economic maximization.[1] One might argue that the pricing behavior of a cartel such as OPEC is more the result of evolving political factors, but in fact most cartels set their prices over time in ways that tend to maximize *economic* objectives. So far OPEC has done more or less the same thing, and to the extent that it continues to do so, economic maximization should provide a reasonable basis for explaining and forecasting the price of oil.

---

*Robert S. Pindyck earned his Ph.D. in economics from the same institution where he is now Associate Professor in the Sloan School of Management, Massachusetts Institute of Technology. He has been consultant to or research advisor with: Dyanamics Associates, Inc.; International Institute for Quantitative Economics, Montreal; Ministry of Planning, Republic of Tunisia; Federal Reserve Board of Governors, Washington, D.C.; Institute for Applied Economic Research, Montreal; National Bureau of Economic Research; Federal Energy Administration; and Development Research Center of the World Bank, Washington, D.C. He is author or editor of five volumes, including *The Economics of the Natural Gas Shortage: 1960-1980* (North-Holland, 1975) and the two-volume *Advances in the Economics of Energy and Resources* (1978). Professor Pindyck's articles have appeared in *Econometrica, Journal of Development Economics, Foreign Policy*, and *Bell Journal of Economics and Management Science*, among others. He serves on the Advisory Board of Editors of the *Journal of Energy and Development*. The research leading to this paper was conducted as part of the M.I.T. World Oil Project and was funded by the RANN Division of the National Science Foundation under Grant no. DAR78-19044. The author wishes to acknowledge his appreciation to the National Science Foundation and to Tom Cauchois for his excellent research assistance.

[1] R.S. Pindyck and E. Hnyilicza, "Pricing Policies of a Two-Part Exhaustible Resource Cartel: The Case of OPEC," *European Economic Review*, August 1976, pp. 139-54; R.S. Pindyck, "Gains to Producers from the Cartelization of Exhaustible Resources," *Review of Economics and Statistics*, May 1978, pp. 238-51 and "OPEC's Threat to the West," *Foreign Policy*, spring 1978, pp. 36-52.

I have calculated optimal pricing policies for OPEC using a small model that quantitatively describes the characteristics of total world oil demand (and its response to income growth and price changes), non-OPEC oil supply (and its response to price changes as well as resource depletion over time), resource depletion within OPEC, and the different levels of reserves and different rates of time discounting among OPEC members. The model calculates a price trajectory for OPEC that maximizes its sum of present and future discounted profits. Using the model, we found OPEC's best price for 1978 to be about $12/barrel in 1975 dollars, or some $14-15/barrel in 1978 dollars — just slightly above the actual price. Furthermore, in real terms, this price should grow by only about 2 percent per year over the next 10 years, and slightly faster over the following 10 or 15 years. A number of sensitivity studies indicated that these results are rather robust with respect to assumptions about the characteristics of demand and supply response, the choice of discount rates, and the like.

Given these results, it would appear that OPEC's pricing problem has a reasonably clear solution, at least for the short term. The price of oil should rise only slightly in real terms, and given the rates of inflation that have prevailed over the past year, together with the changes that have taken place in the relative values of the dollar and several other key currencies, this means a rate of *nominal* price increases in the vicinity of 10 percent per year. In fact, this will effectively be the rate of growth of price over the coming year as a result of the recent OPEC meeting (1978) in Abu Dhabi.

There are now a number of problems, however, that make OPEC's optimal price in the longer term somewhat uncertain. Two problems in particular are causing growing concern to cartel members. The first is over the allocation of production cutbacks: to what extent may Saudi Arabia and a few other producers have to carry a growing share of these cutbacks, and at what cost? The second problem is the threat raised by the possibility of large-scale Mexican oil production: if Mexico becomes a major producer over the next five to 10 years, to what extent would this erode OPEC's monopoly price? The answers to these questions are important determinants of OPEC's ability to maintain its monopoly price over the longer term and are therefore of interest to oil-importing countries like the United States as well as the cartel itself.

Saudi Arabia, Kuwait, and the United Arab Emirates have indeed endured the largest percentage of production cutbacks (they have recently been producing at about 60 percent of capacity), but other OPEC members have also cut back their production. So far this year (1978), Libya, Iraq, and Nigeria have been producing at about 75 percent of their maximum sustainable capacities, Iran — before the strikes that shut down the oil fields — was producing at about 80 percent of capacity, while production of other OPEC members has ranged between 85 and 100 percent of capacity.

There is some question, however, as to how many OPEC members will be able (or willing) to hold production down in the future. Although OPEC as a whole has accumulated a total surplus of some $170 billion since 1973, only five of the 13 members — Saudi Arabia, Kuwait, the Emirates, Qatar, and Libya — are now running annual surpluses. The other members are dipping into their reserves to finance import needs, or, as in the case of Iran, borrowing heavily from abroad.

Furthermore, the squeeze is likely to become much greater in the years ahead. Countries such as Iran, Algeria, Indonesia, and Venezuela launched development plans and made import commitments on the assumption of a sustained level of oil revenues. But OPEC revenues will fall as world oil demand and non-OPEC supplies continue to adjust to the price increases that have already occurred a few years ago. As a result, OPEC's cash-hungry members may be even less willing to produce below capacity in the future.

If Saudi Arabia, Kuwait, and the United Arab Emirates are left alone to maintain a cartel price, their economic interests would be best served by a *reduction* in that price (a reduction which would further squeeze the cash-hungry producers, making them still less willing to absorb cutbacks). At issue is the extent to which the monopoly price would fall.

Another potential problem for OPEC is the possible emergence of Mexico as a major oil producer. Although there have been conflicting estimates, it now appears that Mexico has potential recoverable oil reserves of 150 to 300 billion barrels. (Saudi Arabia has proved reserves of 150 billion barrels and potential reserves of about 500 billion barrels.) Mexico's production today is just over 1 million barrels per day (b/d), but is expected to increase to over 2 million b/d by 1980. It is conceivable that by 1987 or 1988 Mexican production could reach 7 or 8 million b/d, about equal to current Saudi production.

A major increase in Mexican production would reduce the net demand for OPEC oil, and thereby reduce the revenue-maximizing price that OPEC could set. The question is *how much* it might reduce the price. Can the importing countries look to Mexico to return them to an era of cheap oil?

We have used our model of optimal OPEC oil pricing to measure roughly the possible impact of a shifting burden of production cutbacks to the Saudis and a few other countries and of increases in Mexican production. We find that the problem of allocating production cutbacks is likely to be the most serious threat to OPEC's ability to maintain a high price. A high level of Mexican production also would serve to depress the monopoly price, but not by a very large amount. However, Mexican production would significantly erode the flow of OPEC profits.

In the next section we briefly review the structure of our optimal OPEC pricing model. We also present the optimal price and production trajectories obtained previously from that model, since they will serve as a "base case" for determining the impacts of a reallocation of production cutbacks and an increase in Mexican production. Next, we use the model to examine the longer-term issues just raised. We conclude with some remarks about the evolution of world oil markets, and the implications for United States policy.

The Calculation of Optimal Prices for OPEC

OPEC's economic interests are best served by adjusting the price of oil so that it always maximizes the flow over time of all current and discounted future revenues, i.e., so that it maximizes the equity value of OPEC's oil reserves. There are a number of important issues that affect the determination of OPEC's optimal price. First, the demand for cartel oil is "residual" demand, that is, the difference between total (non-communist) world demand for oil and the supply of oil from non-cartel countries. This residual demand is particularly sensitive to price, since increases in price tend both to decrease total demand as well as increase non-cartel supply.[2] Of course, the impact of a price change occurs only slowly, since total demand and non-cartel supply depend on stocks of energy-consuming and energy-producing capital that cannot be converted or shifted rapidly. This would suggest that OPEC might try to make a "quick killing" by steeply increasing the price, taking advantage of the fact that the demand for its oil would fall only after a delay of several years. Such a strategy could indeed work, but only once, and at that only in moving from a period in which the world had been used to low oil prices to a period of high oil prices, i.e., it could work (and did work) in 1973-1974. In fact, it explains why the price of oil has dropped in *real* terms from 1974 to 1978.[3]

Another important consideration is the fact that OPEC's oil resources are finite and will eventually run out. This does not mean, however, that OPEC should sharply increase price and cut production in order to conserve its resources for the future. Over-conservation is as bad as under-conservation, since it simply reduces the net value of a resource to its owner. OPEC's

---

[2]For a discussion of the characteristics of world oil demand and world energy demand in general, see R.S. Pindyck, *The Structure of World Energy Demand* (Cambridge, Massachusetts: Massachusetts Institute of Technology Press, 1979).

[3]It is important to stress that such a "quick killing" strategy would not work now with the world in a high oil price equilibrium. Sharply increasing the price of oil again would certainly increase short-term revenues, but it would diminish future revenues much more, and therefore would be undesirable to OPEC members (unless they are extremely shortsighted, and there is no evidence to indicate that this is the case).

problem is to find a way to exploit its resources so as to balance the revenue obtained today from current production with the (discounted) revenues that could be obtained from future production. This, in fact, is just the problem facing any producer of an exhaustible resource, and its solution usually calls for smooth and gradual changes in price.

These considerations were incorporated into a dynamic optimal pricing model for a monopolistic oil cartel. We review that model here, since it provides the basis for our analysis in this paper.

The world oil market is described by the following equations, all of which were parameterized to be consistent with the reserve, production, and elasticity estimates of the OECD, and with average elasticity estimates obtained from aggregate time series data:

$$TD_t = 1.0 - 0.13P_t + 0.87TD_{t-1} + 2.3(1.015)^t, \qquad (1)$$

$$S_t = (1.1+0.10P_t) \cdot (1.02)^{-CS/7} + 0.75S_{t-1}, \qquad (2)$$

$$CS_t = CS_{t-1} + S_t, \qquad (3)$$

$$D_t = TD_t - S_t, \qquad (4)$$

$$R_t = R_{t-1} - D_t, \qquad (5)$$

TD_t = total demand for oil (billions of barrels per year — bb/yr),
D_t = demand for cartel oil (bb/yr),
S_t = supply of competitive fringe (bb/yr),
CS_t = cumulative supply of competitive fringe (billion barrels — bb),
R_t = reserves of cartel (bb),
P_t = price of oil ($ per barrel), in real 1975 dollars.

The demand equation (1) is based on a total demand of 18 billion barrels per year (bb/yr) at a price of $6 per barrel, and at that price the short-run and long-run price elasticities are 0.04 and 0.44, respectively (with a Koyck adjustment), while at a $12 price the elasticities are 0.09 and 0.90, respectively. The last term in the equation provides an autonomous rate of growth in demand of 1.5 percent per year, corresponding to a long-run income elasticity of 0.5 and a 3 percent real rate of growth in income. Equation (2) determines supply for the competitive fringe and is based on a level of 6.5 billion barrels per year at a $6 price. The short-run and long-run price elasticities are 0.09 and 0.35, respectively, at the $6 price, and 0.16 and 0.52, respectively, at a $12 price. Depletion of competitive fringe reserves pushes the supply function to the left over time. After a cumulative production of 210 billion barrels (e.g., 7 billion b/yr for 30 years) supply would fall (assuming a fixed price) to 55 percent of its original value.[4]

The objective of the monopolistic cartel is to maximize the sum of discounted profits. Here the average cost of production for the cartel rises

$$\underset{\{P_t\}}{\text{Max}}\ W = \sum_{t=1}^{N} \frac{1}{(1+\delta)^t} [P_t - 250/R_t] D_t. \qquad (6)$$

hyperbolically as R_t goes to 0. The initial reserve level is taken to be 500 billion barrels, and initial average cost is 50 cents per barrel. The planning horizon N is chosen to be large enough to approximate the infinite-horizon problem (N = 40 years usually provides a close enough approximation).

Since average costs become infinite as R_t approaches 0, a resource exhaustion constraint need not be introduced explicitly, so that equations (1) to (6) represent a classical, unconstrained, discrete-time optimal control problem, and numerical solutions are easily obtained. The optimal price trajectory for a discount rate of .05, OPEC production, OPEC reserves, and discounted profits are given in table 1.

Table 1

BASE CASE SOLUTION

t	P (in 1975 dollars)	TD	D	R	π_d
1975	13.24	17.24	9.94	488.5	126.5
1976	11.19	16.88	9.23	478.6	93.8
1977	10.26	16.72	8.94	469.3	78.9
1978	9.90	16.66	8.87	460.4	71.7
1979	9.82	16.66	8.91	451.5	67.9
1980	9.88	16.69	9.00	442.6	65.7
1985	10.84	16.96	9.67	396.3	60.6
1990	11.98	17.32	10.40	346.5	56.3
1995	13.18	17.74	11.15	293.0	51.8
2000	14.46	18.22	11.91	235.7	47.1
2005	15.92	18.75	12.66	174.6	42.5
2010	20.29	18.67	12.55	110.5	41.0

Observe that the optimal monopoly price is \$13 to \$14 in the first year (1975), declines over the next five years to around \$10, and then rises

[4]There is no fixed upper bound on cumulative production by competitive fringe countries; there is always *some* price at which additional supplies would be forthcoming. For example, after 210 billion barrels have been produced, a price of \$18.50 would be needed to maintain production at 6.5 billion b/yr.

slowly. This price pattern is a characteristic result of incorporating adjustment lags in the model — it was optimal for OPEC to charge a higher price initially, taking advantage of the fact that net demand can adjust only slowly. Of course, these results are dependent on the particular model and parameter values described above. However, changing the model's parameters has only a small effect on the numerical results. For example, if the elasticities (short- and long-term) of total demand are *doubled*, optimal prices decrease by less than 20 percent. Doubling the elasticity of competitive supply results in a decrease in price of about 10 percent. Replacing the total demand and competitive supply equations with isoelastic equations (using the $6 elasticities from the linear equations) results in price trajectories that are within 15 percent of those reported in table 1. Finally, doubling or halving initial OPEC production costs, or changing the initial level of OPEC reserves from 500 billion barrels to 800 billion barrels has little effect (less than 10 percent) on the optimal price trajectory.

Output Allocations, Mexican Oil, and the Optimal Cartel Price

In table 2 we show the proved reserves, production capacity, and actual production level for each of the OPEC countries. The countries are divided into two groups, the first of which consists of "saver" countries with less immediate need for cash, and the second consists of "spender" nations with greater immediate needs for cash. Note that the "saver" countries also have much higher reserve levels; the average reserve production ratio for the first group of countries is 54.2, while it is only 26.1 for the second group. Note also that the largest production cutbacks have been absorbed by the first group of countries. In 1978 the average rate of capacity utilization was .67 for the first group, and .82 for the second group.

It is important to stress that although capacity utilization is higher in the second group of countries, it is still well below 1, and this has contributed to the ability of the cartel to maintain recent prices. We now would like to determine how the optimal cartel price would change if only Saudi Arabia and perhaps a few other countries had to bear the burden of all production cutbacks.

To do this, we assume, first, that the cartel consists only of Saudi Arabia, Kuwait, Qatar, and the Emirates (Abu Dhabi and Dubai). The other OPEC countries are assumed to produce at capacity (independent of price), which is taken to decrease from its actual 1978 level at a rate of depletion equivalent to that in the competitive fringe countries. In other words, the depletion factor that was introduced in equation (2) for the competitive fringe also is applied to the exogenous production level of these other OPEC members. The optimal pricing decision therefore resides with Saudi Arabia and the other "core" countries just mentioned, given the exogenous production of the other OPEC countries and given the supply function for the competitive fringe.

Table 2

OPEC CAPACITY AND PRODUCTION
(in billions of barrels)

	Reserves[a]		Capacity[b]	Production[c]		
	1975	1978	1978	1975	1977	1978
"Savers"						
Abu Dhabi	29.5	31.0	.95[d]	.51	.61	.53
Dubai	1.3	1.4		.09	.12	.13
Iraq	34.3	34.5	1.17	.82	.81	.88
Kuwait	68.0	67.0	1.10	.75	.65	.62
Neutral Zone	6.4	6.2	.26	.16	.13	.14
Qatar	5.8	5.6	.26	.16	.16	.17
Saudi Arabia	148.6	150.0	4.20	2.58	3.29	2.74
Libya	26.1	25.0	.91	.54	.75	.69
Total	320.3	321.0	8.85	5.63	6.54	5.92
% increase		--			16%	-9.5%
"Spenders"						
Iran	64.5	62.0	2.55	1.95	2.07	2.05
Algeria	7.4	6.6	.44	.35	.34	.37
Gabon	2.2	2.1	.11	.07	.08	.08
Nigeria	20.2	18.7	.88	.65	.77	.62
Indonesia	14.0	10.0	.66	.48	.62	.61
Ecuador	2.4	1.6	.07	.06	.07	.07
Venezuela	17.7	18.2	.88	.86	.82	.76
Total	128.4	119.2	5.59	4.42	4.77	4.56
% increase		-7.1%			7.9%	-4.4%

[a]Reserves as of 1/1/76 and 1/1/78, *Oil and Gas Journal.*
[b]Capacity as of 9/78, *Petroleum Intelligence Weekly.*
[c]Production in yearly amounts, 1978 annualized from seven months, *Oil and Gas Journal.*
[d]Includes Bahrain, which is not a member of OPEC.

To take this one step further, we also examine the implications of Saudi Arabia absorbing all production cutbacks by itself. Now every OPEC country except for Saudi Arabia produces at capacity, with exogenous production again declining at the same rate of depletion.

Next, we wish to determine how a significant increase in Mexican oil production over the next several years might affect OPEC's optimal price. To do this we once again include all of OPEC's members in the cartel and assume that it behaves as a monopolist as in the base case. Here, however, we include an additional source of supply in computing the cartel's net demand for oil, replacing equation (4) with $D_t = TD_t - S_t - SMEX_t$ where SMEX is an exogenous variable representing Mexican production. We assume that Mexican production rises yearly from 1 million barrels per day in 1975 to 7 million barrels per day (2.55 billion barrels per year) in 1985. From 1985 onwards we assume that Mexican production declines with the same rate of depletion as in the competitive fringe. Thus, Mexican production falls to 2.26 billion barrels per year in 1990, and 1.82 billion barrels per year in the year 2000.

The optimal price trajectories for these scenarios and the base case are shown in figure 1. Observe that if Saudi Arabia alone or Saudi Arabia together with a group of core countries are left to bear the full burden of production cutbacks, the optimal price is significantly lower than it would be otherwise, particularly during the first 15 or 20 years. In 1985, for example, the price is about 20 percent below that in the base case. The presence of Mexico also tends to reduce the optimal cartel price, but not by as much (less than 10 percent), and only after 1980. It is interesting to note that the optimal cartel price is actually ***higher*** in the first one or two years if an increase in Mexican production is anticipated. The reason for this is that the incentive to the cartel for a "quick killing" is greater since the growth in Mexican production means that there is less to lose later on from a resulting decrease in world oil demand and increase in competitive fringe production.

Figure 2 shows total OPEC production for the base case and for the cases in which (a) Saudi Arabia alone and (b) Saudi Arabia, the Emirates, and Kuwait bear the burden of production cutbacks. Observe that when Saudi Arabia and the Emirates bear the burden of production cutbacks, cartel production is significantly larger, which is consistent with the much lower price. However, this increase in production is not solely due to the higher production levels of the non-participating OPEC members; surprisingly, Saudi Arabia's production is also higher when it is left alone to do all of the cutting back. The reason for this is that the net demand curve for Saudi Arabia is now more elastic over a wider range of price, so that profits are maximized by an increase in production.

In figure 3, the present discounted value of profits is shown for the base case and the case with Mexican production increasing to 7 million barrels per day by 1985. Note that although the price trajectories in the two cases are not very different, there is a significant difference in OPEC's profits. With Mexico on stream, the high monopoly price is maintained, but OPEC has a smaller share of world production. In fact, by summing the present discounted value of profits in the two cases, we can determine the cost to OPEC of Mexican production. Summing to the year 2005, we find that the net present value of OPEC's profits without Mexican production is $1,843.5 billion, while the Mexico on stream, the net present value is $1,590.8

billion. In net present value terms, the cost to OPEC of Mexican production is therefore about $253 billion.

Figure 1

OPTIMAL PRICE TRAJECTORIES

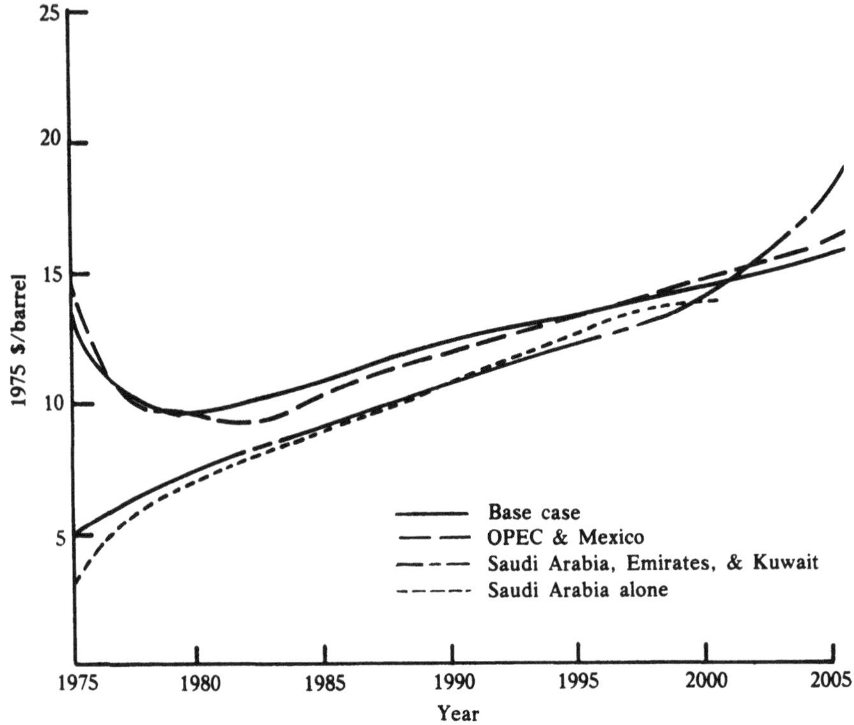

We thus find that if OPEC cannot find a means of distributing production cutbacks, we could expect to observe a significant decline in the cartel price. The effect is large because with only Saudi Arabia and a few other core producers absorbing production cutbacks, cartel control over the market is greatly reduced. As far as Mexican oil is concerned, we find that if OPEC can successfully distribute production cutbacks among all of its members, it could limit the impact of Mexican production on the monopoly price. Even if Mexico increased its production to 7 million barrels per day as early as 1985, the monopoly price would only be reduced by some 8 to 10 percent at most. On the other hand, OPEC's share of the oil market would be reduced, and the PDV (present discounted value) of its profit flow would fall by some 10 to 12 percent.

Figure 2

OPEC PRODUCTION

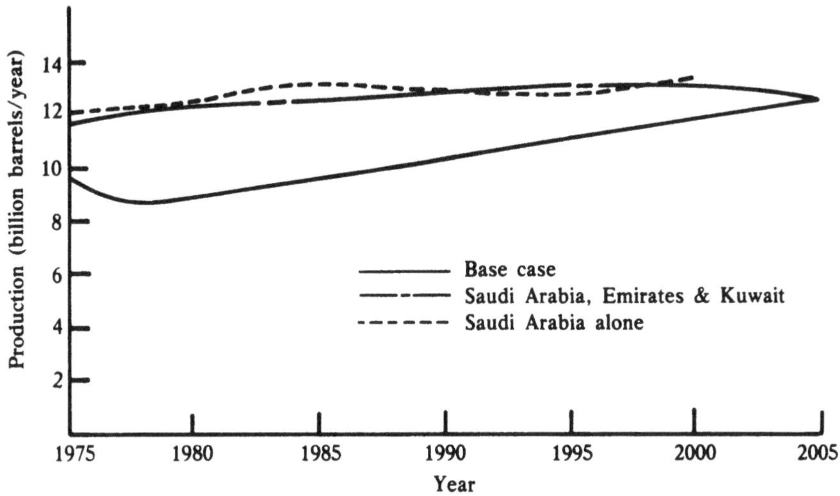

World Energy Markets and Energy Policy in the Importing Countries

Considerable attention has recently been focused on reports by the Central Intelligence Agency (CIA) and by the Workshop on Alternative Energy Strategies (WAES) which predict a crisis in which energy demand will exceed supply beginning in the 1980s.[5] According to these reports, this crisis will occur because the demand for OPEC oil will grow steadily as world oil demand increases faster than non-OPEC oil supply. OPEC production capacity, on the other hand, will remain fixed or even gradually fall, so that eventually there will be excess demand for OPEC oil. According to the reports, the crisis begins at this point; there is a shortage of oil, and what oil is available is sold at prices that increase suddenly and dramatically.

This scenario is grossly unrealistic, and ignores two important facts. First, the demand for OPEC oil (and to a lesser extent OPEC production capacity) is highly sensitive to the price of oil. Second, OPEC will gradually and steadily increase the price of oil over time, and at such a rate that capacity constraints will never become binding.

OPEC is now operating well below its production capacity and has

---

[5]United States, Central Intelligence Agency, *The International Energy Situation: Outlook to 1985*, ER 77-10240U (April 1977) and Workshop on Alternative Energy Strategies, *Energy: Global Prospects 1985-2000* (New York: McGraw-Hill Book Company, 1977).

considerable room for capacity expansion in the future. OPEC production in 1977, for example, was about 29 or 30 million barrels per day (b/d), while capacity was over 35 million b/d. While the production capacity of some low-reserve countries such as Iran and Venezuela will decline during the 1980s and 1990s as reserves are depleted, the production capacities of such high-reserve countries as Saudi Arabia, Kuwait, and the Emirates can be expanded considerably. In fact, Saudi Arabia and some of her high-reserve neighbors could as much as double their capacity, *particularly if there is a revenue incentive to do so.* And, as long as these countries are concerned with the pursuit of their economic interests, they should increase their capacity and exploit their oil resources at the optimal price. To do otherwise would be to waste the economic value of their resources.

Figure 3

PROFITS WITH AND WITHOUT MEXICO

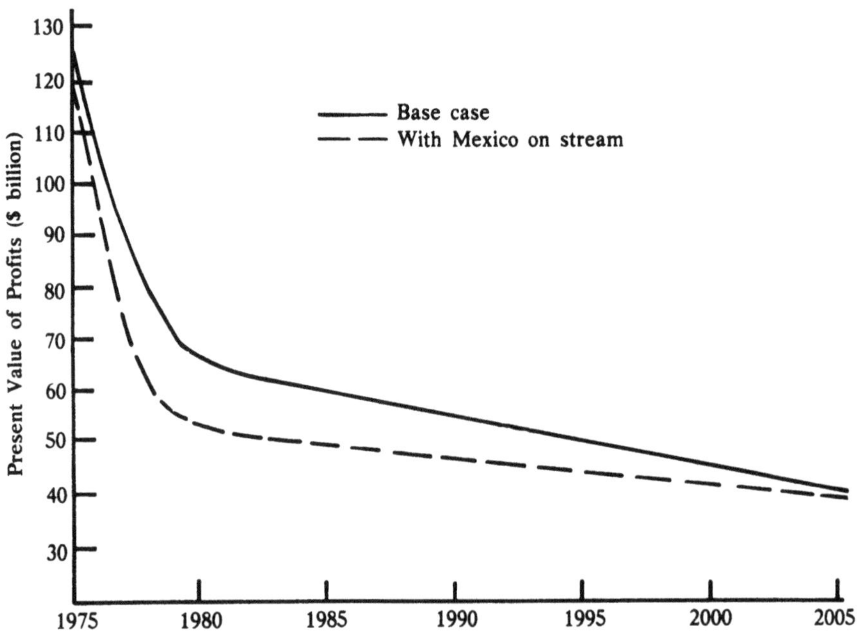

We thus should expect a gradual evolution of the world oil market, so that the price of oil slowly increases at rates that will not have any serious impact on the macroeconomic output of most of the industrialized countries. At the same time, the production capacities of some OPEC countries will increase enough to more than offset possible declines in the capacities of such countries as Iran and Venezuela. In addition, gradual increases in the price of oil will serve both to dampen the growth in world oil demand and to stimulate the production of oil from non-OPEC sources. Although oil will become gradually more expensive, there is no reason to expect the sort of price and output scenario that the CIA and WAES have predicted.

Although the CIA's scenario is extremely unrealistic, those countries that are heavily dependent on imports of OPEC oil (the United States included) must be concerned with the possibility of a sudden cutback in OPEC production and run-up in price. Unforeseen political events, such as those we have witnessed in Iran, could result in a sudden change in OPEC production and in the price of oil. It is important to stress, however, that such a crisis could occur as easily now as in the mid- or late 1980s, and it is not this kind of crisis that has been of concern to the CIA or the Carter Administration. The possibility of such a crisis should be of greater concern to the administration; it is an important reason for pursuing a domestic energy policy directed at reducing or at least limiting oil imports. A production cutback could occur, and the United States, as well as the other industrialized importing countries, should take measures to reduce their potential vulnerability.

A significant measure is the development of strategic oil reserves. The United States is presently beginning to implement a program to store oil in underground salt domes near the Gulf Coast. Unfortunately, the program is proceeding slowly, and the plan to reach the one billion barrel target in 1984 may never be reached. Furthermore, should an embargo or production cutback occur within the next few years, the United States could expect little relief from the stockpile. It therefore would be wise to speed up, and perhaps expand, the stockpile program. And, comparable stockpiles also should be developed by the other import-dependent countries.

In the case of the United States, it is also critical to recognize that our growing dependence on imports to a large extent has been the result of our policies of price controls on natural gas and the price controls-entitlement system for crude oil. Energy prices should be deregulated in the United States, and the entitlements system should be disbanded altogether. Letting energy prices in the United States rise to world levels will help considerably to reduce our dependence on imports, and it will permit domestic energy markets to operate efficiently, thereby reducing the total cost of energy in the long run.

The United States should also be doing whatever it can to encourage greater Mexican energy production. As we have seen, the impact of increased Mexican production on OPEC's optimal price is not likely to be large, but it would certainly be noticeable, and a high level of Mexican output is desirable for other reasons as well. Mexico would be a much more secure source of supply for the United States, and a source of supply less likely to require political payments in addition to economic ones. The

Department of Energy's recent veto of the United States-Mexican natural gas deal was unfortunate indeed.

The importing countries can hope that OPEC will fail to solve its long-run problems, but they had better not count on this. Mexican oil is still largely speculative, and the continuing turbulence in Iran might lead to a large and permanent drop in oil production, which could by itself solve the allocation problem for the cartel. Even if Iran's production stabilizes at its earlier levels, some of the non-core countries will lose revenues if they fail to hold back production. These countries may well have the foresight and discipline to look out for their economic interests and hold the line on production. It is therefore likely that oil will become increasingly expensive in the years ahead.

4

OPTIMUM PRODUCTION AND PRICING POLICIES

*Fadhil Al-Chalabi and Adnan Al-Janabi**

Price Formation in the Oil Industry

Historically, the only semblance of a "market" in the oil industry could be found in the early days of competition within the United States, before the rise of a monopoly for buying and distributing oil from the many producers. Early in this century, developing producing countries were given nominal fixed payments for oil produced and exported. Even with the rise of the so-called profit-sharing system, oil was produced, transported, refined, and distributed as products through the integrated channels of the major oil companies.

*Fadhil Jafaar Al-Chalabi, currently Deputy Secretary General of OPEC, holds a Doctorat d'état from the University of Paris in economics as well as degrees from the University of Poitiers and Baghdad University. Prior to assuming his present post, Dr. Al-Chalabi was the Assistant Secretary General of the Organization of the Arab Petroleum Exporting Countries (OAPEC) and had served as well as both the Permanent State Under-Secretary of the Ministry of Oil and the Director General of Oil Affairs in that ministry in Iraq and as Economic Affairs Officer with the United National Conference on Trade and Development (UNCTAD) in Geneva. The author has been a member of: the Board of Directors of the Iraq National Oil Company (1968-1976); the Government Higher Commission for Marketing of Iraqi Oil (1973-1976); and the Central Bank of Iraq. Adnan Al-Janabi, presently Head of the Economics and Finance Department of OPEC, holds degrees in economics from the Universities of London and Loughborough, England. Prior to joining the OPEC staff, the author served in the Economics Department of the Iraq National Oil Company as well as holding a teaching appointment in petroleum economics at Baghdad University. This paper draws heavily on previous research and published work by the coauthors including: F. Al-Chalabi, "The Administrable Nature of Pricing OPEC Oil," *OPEC Review*, September 1978 and "The Middle East Crude Oil Availability," a paper to the Sixth Energy Policy and Decision-Making Seminar, Sanderstolen, Norway, February 1979; A. Al-Janabi, "Determinants of Long-Term Demand for OPEC Oil," *Journal of Energy and Development*, spring 1978, pp. 347-65 and "Production and Depletion Policies in OPEC," paper presented to the University of Surrey, England, January 1979. The views expressed in this article are those of the authors.

The horizontal integration of the oil industry upstream and its vertical integration of each company rendered the international market for crude oil exports from OPEC states a totally administered one. The major oil companies between them acted as a cartel in their various operations, including price setting. The transfer of oil within the integrated channels of each company individually and between the members of the group did not represent anything more than a nominal exchange for accounting purposes. Consequently, throughout that period a market for crude oil did not exist in the sense of a free interaction of supply and demand.

The companies' price administration (setting unilaterally the level of crude oil posted prices) served the purpose of determining the "tax-paid cost" of lifting crude oil from the producing areas. It was that latter cost, and not the "posted price," which represented the real cost of oil acquisition by the consuming countries, or the net transfer of wealth from the producers to the consumers. The companies' administration of that transfer helped to "optimize" the energy cost for the postwar reconstruction of the developed economies of Western Europe and Japan.

Various changes which took place in the posted prices of Middle Eastern oil before, during, and after World War II indicate the degree to which price formation was an administrative process rather than one generated by market forces. The outcome of the various changes in posted prices of Middle Eastern oil, calculated on the basis of so-called "basing points" and phantom transportation costs, was that by the beginning of the 1960s the f.o.b. prices of Middle Eastern oil were brought down to almost half of their equivalent crudes f.o.b. at the United States East Coast, from an earlier level when they were higher than United States prices before the war. The apparent link with United States oil prices had no real meaning throughout this period since American oil never competed in reality with Middle Eastern oil, especially as the American markets were increasingly isolated from the international oil market by administrative arrangements.

With the rise of OPEC a regime of coadministration of prices came about, putting an end to the unilateral role of the international majors in setting posted prices for crude oil entering international trade. Despite the operations of a small, arms-length market for crude oil sold to third parties due to the weakening of the grip of the international majors on oil production, the bulk of internationally traded oil continued to be handled through the integrated channels of the majors. This remained the most important element in price formation during the sixties. The freezing of posted prices throughout this period had the effect of strengthening the market price structure by setting a fixed floor for third-party transactions, since the tax-paid cost of oil lifted by the majors represented a minimum to which such "realized prices" could be transacted. The prices and fiscal system brought about by the Tehran Agreement of 1971 strongly asserted

OPEC's role as a price coadministrator. According to the Tehran and subsequent related agreements, realized prices as distinct from posted prices had to increase annually by a minimum amount equivalent to the scheduled increases in tax-paid cost.

With the take-over of oil pricing responsibility by OPEC since October 1973, the administration of prices reverted totally to the exporting countries themselves. During the past five years, OPEC has managed to assert its role as price setter during different phases in the oil market. It managed to adjust prices upwards by 10 percent during 1975 despite the record drop in international demand for oil. It has been demonstrated conclusively that demand for oil is much more related to the level of economic activity in the main consuming nations than to changes in the f.o.b. prices of crude oil. This fact, combined with the past and present structure of the oil industry, makes it possible for the administration of oil prices to be effectively undertaken whether with a concession regime through foreign oil companies or under the present take-over of effective control of their natural resources by the exporting countries.

The administered nature of crude oil prices should not be construed to mean that OPEC (and previously the companies) could act without regard to other forces interested in the price issue. In fact, crude oil prices can be viewed as a result of the interaction of three major forces: (1) the consuming governments, which are trying to minimize foreign exchange costs of imported energy; (2) the oil companies, trying to maximize corporate profit through minimizing total lifting costs; and (3) the producing countries, trying to maximize the real transfer of wealth arising from oil exports. These main pressures continue to determine oil prices through a process of resolving conflicting interests by means other than the free interplay of market forces.

A market exists (with constraints) for final consumers who use petroleum products in light of their income, the price of specific petroleum product, and the availability and price of alternatives. To the final consumer the cost of a composite barrel of oil was about \$12 to \$13 per barrel in Europe in the early 1960s, when OPEC members were getting \$ 0.70 to \$ 0.90 per barrel. How the value (or total rent) of a barrel of oil was (and still is) divided, i.e., which price tag is set on the internationally traded barrel of crude oil, remains to be determined by the result of the relative power of the three main conflicting interests. Since 1973, the relative strength of the conflicting forces has shifted to the extent that the oil companies are less influential than the producing countries in determining the split of oil rent.

This trend has been strengthened by the following tendencies: (1) the consuming countries consider energy imports a matter of national security; (2) the producing nations have general control over domestic oil operations; and (3) internal energy demand in the consuming countries (investments, substitutions, mandatory conservation, foreign exchange allocations for energy imports, and the like) are areas of government policy rather than of market equilibria.

The energy policies of consuming nations which are net importers of petroleum tend to aim at minimizing such net imports regardless of the relative prices of available alternatives. In the area of electricity generation where coal, nuclear, and other alternatives can be resorted to, programs to bring in those alternatives at almost any cost are put into motion. In the areas where alternatives are not available or severely constrained, energy policies of consumer governments tend to employ various mechanisms for curbing demand; little attention is paid to the economic effects of such restrictions. This intensive governmental interference in the energy sector makes any assumption based on competitive equilibria practically useless. As a consequence, it is observed that the demand for crude oil is gradually becoming residual within the general determinants of total demand for energy. All economically and technically feasible alternatives are being pushed to their upper physical limits. There is no spare capacity in the real sense of the word in nuclear power stations, hydroelectricity, or coal utilization. Investments in these areas are going forward at maximum feasible speed, allowing for environmental, financial and natural constraints. The only spare capacities available are those related to the production of oil and gas.

Total demand for OPEC oil tends to be a residual one of total world demand for crude oil. Non-OPEC oil coming into the world market from the North Sea and Alaska was developed before the price increases of 1973 and 1974. It could be argued that, as in the case of the coal industry in Europe in the 1960s, oil would still have been produced from those areas even if the costs had been higher than world market prices. Thus, total demand for OPEC oil can be seen as residual demand for world energy apart from direct demand for nonenergy uses. Since all energy sources, other than OPEC oil and gas, are operating at near capacity and investments to extend those supplies are going on regardless of price trends, it could also be argued that demand for OPEC oil is influenced more by physical constraints on alternatives than by prices of crude oil and petroleum products.

The energy policies of the consuming countries are based on estimates of the level of oil imports needed to meet their energy demand without any meaningful reference to the relative prices of various energy sources. For example, the IEA (International Energy Agency of the OECD) objective of maintaining an import level of 26 million barrels of oil per day by 1985 is described as necessary to "maintain an equilibrium in market conditions" without much reference to the price mechanism. It is important to emphasize that demand for OPEC oil in the short and medium term would not be determined in the light of crude oil prices, but rather as the minimum physically necessary to fill the energy gap of the consuming countries.

Crude oil prices could be increased by the same amount now taken by consuming governments as tariffs and taxes on crude oil and products without any effect on demand. They could be increased by an amount equal to the subsidies, rebates, and the like now being allowed to competitive sources such as coal and nuclear energy, if oil were allowed in on a fully competitive basis. This increase would again be absorbed without much effect on the final consumer. It also seems justifiable to say that in the short and medium term crude oil prices could be increased substantially above effective parity with the cost of incremental alternatives since there is no way left to reduce demand apart from the general slowing down of economic activity.

In view of the high degree of intervention in the international trade of crude oil by producing governments, consuming nations, and companies, and given the role of energy policies in replacing the internal energy "markets" of consumer countries, it is important for the producing nations to look for an alternative mechanism of price formation. Here we are basically referring to the pricing of crude oil in international trade. OPEC is already involved to a certain degree in the planning of oil prices. A close examination of the oil industry in the context of energy supply and demand shows that the planning of oil prices is not only feasible but necessary.

The depleting nature of hydrocarbons and the very long lead time involved in energy development make price planning a necessary prerequisite for a smoother transition to viable alternatives in the future.

Parameters for Oil Price Planning

Demand for crude oil is derived from the demand for petroleum products. Crude oil is only consumed directly in very few cases, notably in power generation. In these instances, crude is used without refining as a form of fuel oil. Otherwise, demand for crude oil is derived from the demand for: (1) transportation usage, as gasoline and gasoil (diesel) for internal combustion engines, kerosine for jet turbines, and heavy fuel oil for marine bunkers; (2) direct energy requirements of those products for industrial use, space heating, and household appliances; and (3) nonenergy uses as in the case of chemical feedstocks, lubricants, asphalts, bitumens, and so forth. The only meaningful way to analyze the relationship between prices and demand is to see the effect of product prices on final consumers. The assumption that product prices have a direct relationship to crude oil prices need not be taken for granted.

If we take a closer look at demand decisions and the price levels to which these decisions react, we have to go into greater detail in each category of demand and into the actual price level of each product to the final consumer. In this context, aggregate prices and weighted averages are not very helpful.

The highest prices in all the major consuming countries are charges for gasoline. There are great variations in gasoline prices depending on whether they are considered ex-refinery or at the pumping station. Italy is now possibly the country with the highest gasoline prices, with an ex-refinery price of about $90 per barrel and a tax of $64 per barrel for premium gasoline. All West European governments tend to push gasoline prices up very high indeed through various taxes. Recently, the United States government has indicated its intention to move in the same direction in order to influence demand for gasoline. Demand for gasoline is considered very inelastic; but this is only relative. The new moves to push up gasoline prices even higher are bound to influence life-styles and transportation patterns. This fact is easily seen in the different sizes of cars and the different role of public transport in Europe and the United States. Steeper gasoline prices could reduce gasoline consumption to the point where private transport is minimized and efficiency of passenger miles maximized. Yet, a point will arrive where the demand curve for gasoline in transportation stiffens again due to the lack of alternatives. A certain minimum level of transportation is necessary for normal economic activity, whatever the price.

A similar analysis is applicable to the transportation usages of diesel oil, ATK (aviation turbine kerosine) and bunkering fuel oil. The prices of those products, especially ATK and bunkers, are very much lower than gasoline. But the price elasticity of demand for those products in transportation should not be too much different from that of gasoline. Some reduction in demand could be achieved as prices rise. Initial reductions would represent increased efficiency, elimination of unnecessary consumption, and increased use of capital in the mix of inputs. At a higher level of the price-demand relationship, the demand curve would again tend to be vertical, reflecting the need for minimum air and marine transportation where no serious alternatives to petroleum products exist.

Another use which exhibits little elasticity is the area of nonenergy consumption, especially in chemical industries. Although natural alternatives exist to petrochemicals, those natural alternatives draw on limited and increasingly scarcer natural resources. Even more important is the fact that the cost of hydrocarbon raw materials is relatively low in the total cost of finished synthetic products. This varies from product to product and could reverse if prices of petroleum inputs become excessively high. But within a reasonable range of prices, petrochemical products are influenced more by capital inputs than by raw material inputs.

Lubricants, greases, waxes, and specialized bitumens exhibit similar price-demand patterns. But demand for asphalt used for road surfaces could well be more elastic than other nonenergy uses, due to the high mineral content of the product, as well as the existence of some alternative technologies such as bare-concrete road surfacing. The problem here is the technological difficulty in the medium term of developing refinery processes which leave little bottom residue, if demand for it vanishes gradually. The solution in the short and medium term might very well be to keep the prices of such residues low in order to maintain demand.

The area where demand for petroleum products is fairly elastic is direct open-furnace burning — whether for direct thermal energy or in generating electricity. There are ready or evolving alternatives for most of the uses included in this category. It is also to be noted that energy conversion is least efficient in this classification of uses. Moreover, it is observed that prices of products used in this category are comparatively low for the final consumer, especially in the case of fuel oil prices for power stations. This situation is tolerated by consumer governments despite the waste it entails. Without going into details about the motives behind this apparently irrational pricing policy, we can safely say that crude oil prices can influence demand for these uses more than any other use of petroleum products. The magnitude of demand for domestic heating oil and fuel oil is fairly sizable in Japan and Europe, though not in the United States. Therefore, there is some ground for arguing that if oil prices go higher than those of alternatives available for direct energy and for power generation, demand for nearly half of the oil barrel would be adversely affected. However, recent developments have revealed that there are severe physical constraints on the ability of alternatives to compensate for petroleum products even in the most readily applicable use in power stations. These physical constraints push the demand curve for petroleum products sharply upwards, making it inelastic even for the uses which were thought to be most responsive to price changes. But it may be argued that in the long run it is such uses, which are both inefficient and replaceable, that may react most to steep price increases.

Price elasticity could be effective if price rises were high enough to generate confidence in investment in substitutes, and, more important, if they lead consuming governments to adopt policies encouraging conversion and substitution.

Income elasticity of demand for energy in general, as well as for most specific uses, is very high. Economic growth has a direct bearing on energy consumption. In most cases, income elasticity is close to unity. In light of conservation measures in the industrialized countries, it is gradually moving below 1. Some expect income elasticity of demand for energy to go down to over 0.8 during the next decade in the OECD nations. But no one expects the relationship between total energy requirements (TER) and GNP to weaken drastically. This means that the most critical determinant of demand for energy is the state of the world economy.

In view of the above, pricing policies cannot ignore the effects of prices on the economies of consumer states, neither can prices be immune from the impact of economic changes in those countries.

The first problem to be tackled in the field of oil pricing is the need for a mechanism to adjust these prices in the short term against the effects of inflation and exchange rate fluctuation. It is important when talking about prices to distinguish between nominal and real prices. Despite the recent rises in nominal prices since 1974, real prices have been cut drastically during the past five years, especially due to inflation. The cost of OPEC imports has risen according to the OPEC Import Price Index by about 340 percent since 1973. Even if we take the export price indices of the industrialized nations, which exhibit continued inflation, we would find that the purchasing power of unit oil exports is on the downward trend. The decline of the exchange rate of the dollar during the last two years made this picture look worse, particularly as throughout the seventies the dollar has tended to weaken against the major currencies used by OPEC members for their payments.

An important reality which is often forgotten in analyzing the international oil markets is that the f.o.b. price of crude oil constitutes only one element of the effective price paid by the final consumer. The relationship between the price of petroleum products charged to the final consumers (as distinct from those observed in the so-called open markets) to f.o.b. prices of crude oil is influenced directly by the relative strength of both consuming and producing countries. The OPEC states can only determine what they get for the export of their natural resources, but due to the present structure of the world economy they cannot directly influence the prices charged finally to the consumer. It could be argued that effective oil prices are those of petroleum products, while f.o.b. crude oil prices are only a kind of "tax" segment that is not very different in its significance from the taxes imposed by the consuming countries, whether in the form of import duties, indirect taxes, or income taxes on oil companies. It is obvious that the distribution of the share of each party from total "rent" on petroleum could be shifted to the production side or to the consumption side with limited consequences on the industry itself. It could be argued furthermore that since differential rent arises from the nature of mineral deposits (in the case of mine rent), it should accrue to the resource owner.

Optimal Prices for Exporting Countries

The real price of oil in the long term could be conceptually related to two basic parameters for planning purposes: (1) replacement and substitution costs, and (2) depletion and exhaustion.

The replacement and substitution of oil in thermal uses, especially for the purpose of electricity generation, is already feasible with coal and nuclear energy. But due to long lead times (10 to 15 years in the case of nuclear energy), physical bottlenecks already exist in the supply and usage of alternatives to fuel oil. It is necessary in this case to keep the prices of fuel oil above those of its alternatives for the supply of electricity so that their development would not be damaged by uncertain prospects.

Alternatives to more specific uses of petroleum products, such as transportation, are not available at present. It is feasible to derive gasoline from synthetic crudes (from coal, tar sands, and shale), but neither the costs nor the prospects are clear. It is generally estimated that such alternatives, would cost about twice the present price of oil. But in reality no large-scale investment is going into these alternatives, nor is it envisaged that they will make a significant contribution to energy supply during the remainder of the present century.

In this context, the long-term aim of keeping oil prices in line with alternative sources of energy is not easy to quantify, not only because of the changing nature of "costs," but also due to the elusiveness of the concept itself. Alternatives can only be meaningful if the final use is also identified. In a situation where both costs and uses are constantly changing, it is not clear how one is to quantify the relationship, especially when given the long lead times. Since there is no physical competition, the long-term supply cost of alternatives is too theoretical to serve as a concrete policy instrument.

In the medium term, oil prices could be related to the known supply cost of direct and full replacement of the presently depleted barrel. This could be the highest oil costs, the upper limit of enhanced recovery costs, present solar energy costs, and the like. Theoretically, this would merge with the long-term concept of the "cost of alternative sources."

Apart from the replacement and substitution costs, the problem of depletion and exhaustion has a direct bearing on medium- and long-term pricing strategies. Prices of scarce and fast declining minerals should reflect their reduced availability. In this connection, prices should rise in real terms at a rate reflecting "user cost." By the same token, the time path of oil prices should be such as to encourage its gradual phasing out from less efficient wasteful uses to more efficient uses, especially where no substitutes are readily available. Rational conservation and orderly replacement of scarce resources could only be done with a well-planned pricing strategy.

From the point of view of oil-exporting countries, oil prices must be viewed in terms of the material collateral they get in lieu of the finite resource they deplete. They maximize their gains when they have the greatest access to wealth (best represented by highest economic growth for their economies) at a minimum depletion of oil reserves. In the long term, this means that a production and depletion policy would be judged efficient to the extent that it succeeds in serving the purpose of bringing the economy of the producing country to the highest possible level of economic development sustainable without further need for oil revenues at the time when petroleum is no more available for export. Nominally, this means the highest unit price which earns the foreign exchange necessary for fixed capital formation.

Recent history since 1973 has proved, however, that nominal prices mean very little in themselves. The control of the OECD nations over the supply of equipment, patents, and know-how (as well as markets for potential nonoil exports by OPEC members) means that in a state of economic conflict, cash is not wealth and access to foreign exchange is not a passport to rapid economic development. Exported inflation (coupled with price discrimination) and the decline of the United States dollar have wiped out most of the price adjustments of 1974. Therefore, prices have to be always looked upon in terms of more complete bargains encompassing the totality of relations with the major industrialized countries. The impact of price increases on the other developing nations cannot be ignored either.

On the other hand, conservation policies by producing countries (minimum depletion) cannot be pursued without regard to demand. Both on the pricing side and on the production side, the complexity of relations between producers and consumers has to be taken into account.

It is vital to note that prices have direct implications on production levels. Therefore, planned pricing should be undertaken with complete awareness of the implications of such price on demand. Although it already has been pointed out that in the short term the price elasticity of demand is very low, it has to be recognized, however, that in the medium and long terms significant price increases would eventually lead to higher elasticity, operating not only through the demand schedules of the individual consumers, but also through the policy reactions of the consuming countries, which are probably a more important determinant of aggregate demand for energy and of the process of conversion and replacement by other substitutes. The impact of price increases on economic growth in the consuming countries has a greater immediate effect on total demand since it has been observed that income elasticity of demand is high in all cases. The method of price adjustment is probably also significant since very sudden, large price increases could have a more dramatic impact on economic activity in general and to an even greater degree on the policies of the consumer nations. Gradual price increases, however, might have the effect of being more easily absorbed by the world economy, and they are easier to implement without negative reactions from the consumers, especially if they are carried out in an atmosphere of understanding and cooperation rather than one of confrontation.

Price planning and pricing policies by producers cannot ignore the problems of exhaustion and depletion on the one hand, and of intergenerational equity on the other. Since prices are the main instrument available to producers to influence future output, this should be done cautiously and with full awareness of the impact on future production levels. A price, whatever the concept on which it is based, leading to wasteful demand and early depletion does not reflect successful planning. Conversely, a price which brings demand down too dramatically (by its effect on economic activity, severe conservation, or wide-scale substitution) might also be unwise. A long-term pricing strategy should in itself be a form of a long-term production plan. Naturally, a degree of adjustment in both parameters is inevitable over time. But in all cases, pricing is simultaneously a form of long-term production planning.

A further problem which has a direct bearing on the pricing policies of oil-exporting countries relates to the soaring allocations for further development in the oil sector. The era of cheap oil is over. The development of the Orinoco Belt in Venezuela could tax the country's oil revenues very severely. The same is true of Iran's enhanced recovery program to maintain production levels. Exploration and development efforts in OPEC members are becoming a heavy burden which poses a serious dilemma. Either development of new fields is restricted and declines in old fields are not arrested, or economic development plans are adversely affected by diverting cash from development investment in other high-priority sectors. Naturally, oil-exporting countries would feel that such a burden should be resolved through the price mechanism. This could be done by a pricing mechanism which reflects changes in the marginal supply cost of new oil, or "replacement cost," of substitutes to oil products.

A related problem is the necessity of involving the consuming nations, which are interested in these incremental oil supplies, in the provision of capital funds required for such development plants and in providing the necessary technology for projects which do not contribute to the diversification of the economy but are limited to raw material exports.

This issue is coming increasingly to the forefront due to two significant international developments. First, current account surpluses are disappearing even in countries which had substantial surpluses until recently, with the resultant implication that foreign exchange availability is again becoming a constraint on development expenditure. Second, flexibility in production represented by excess capacity also is gradually going to disappear, especially if conservation-oriented depletion policies are taken into account. This phenomenon generates a rigidity in production levels which, when coupled with the disappearence of current account balances, deprives the oil-exporting countries of any production flexibility in a tight market. It also implies new constraints on their investment priorities. The situation seems to imply that oil prices should increase to the extent necessary for maintaining the present net income from low-cost oil in order to encourage the producers to put up the necessary investments for the higher cost of new reserves. Some international cooperation is called for as well in the area of new oil technology acquisition if a degree of stability is to be achieved in the world oil market in the future.

Production and Depletion Policies

General production and depletion policies in oil-exporting countries are considered efficient to the extent that they are related to the transformation of a depletable comparative advantage into a more permanent comparative advantage in international economic relations. This means that the only rationally acceptable depletion rate is one where the wasting resource is compensated simultaneously and concurrently by an investment process that will yield the highest rate of economic development. The national interest will be optimally met when the resource is depleted at a time when the highest possible level of economic development has been attained. These are the two factors that must be linked: resource depletion and the rate of economic growth. They should be related to an appropriate time horizon, as depletion might be too fast resulting in the disappearance of exports. It should not set in before the highest possible level of economic development has been attained.

In many cases oil revenues are more than the level necessary to cover fixed capital formation in the countries concerned. A closer look at the essential foreign exchange requirements necessary to cover development expenditure would reveal that even lower production levels are currently called for.

It is not practical, in the short term, to advocate a lowering of production levels to the extent that only the basic foreign exchange requirements for development expenditure are covered. However, depletion policies in the medium and long term should aim at avoiding the utilization of oil revenues for immediate consumption. It is important to point out, nonetheless, that there is a great deal of variation between the future needs of nations with high resources and small populations and those with large populations and relatively low reserves. The first group may be able to afford the use of some oil revenues for improving the standard of living of the present generation. The second group of countries, however, is more constrained in its depletion policy, which should aim at more stringent production and depletion practices.

The general trend of this argument would gradually lead to the imposition of increasing constraints on production levels in most oil-exporting countries. This in itself would be an important leverage on pricing policies, which should aim at keeping oil prices at somewhat higher levels than those of the substitutes, in order to allow investment in them. Thus, a long-term balance in the supply and demand of liquid and gaseous hydrocarbons could be achieved.

The oil-producing states are expanding energy-intensive industries. Various projects started since 1974 are just coming on stream; others are in the stage of construction or planning. The choice of energy-intensive technologies in OPEC members is not a passing phenomenon. Under the impact of expected future energy price rises, it is natural that energy-intensive industries and processes are shifted from oil-consuming to oil-producing areas. However, at the moment, the internal prices of petroleum products are very low in many OPEC nations. Nowhere are the prices of these products based on their export value, which leads to wasteful use in some instances.

Consequently, growth rates of energy consumption in OPEC countries have risen significantly since 1972, reflecting higher industrialization levels and the wider dissemination of different energy forms to new segments of the population. This development is taking place in countries with large populations, where population growth rates are some of the highest in the world. Additionally, due to the general development effort, per capita energy utilization is rising. The horizontal and vertical growth of the energy market in OPEC members is bound to continue in the future, with no saturation point in sight. Considering the development plans of OPEC governments and the expressions of intention for further development, very high future energy needs are to be expected. Even if all these aspirations do not fully materialize, the growth rates of domestic energy consumption will still be rather high.

The domestic consumption of petroleum products and the use of gas in various forms will be significant. Except in a few OPEC countries, where production may exceed consumption, most associated gas will be used up internally. In 1977, gas utilization in the OPEC nations reached nearly 2 million barrels per day of oil equivalent. Although historical growth rates of gas utilization have been stable at around 8 percent annually, they are expected to increase significantly in the near future. Gas consumption outside the realm of reinjection for conservation purposes and reservoir pressure maintenance is expected to rise appreciably in an impressive effort to put an early end to the wasteful flaring of associated gas. This trend is strengthened in the light of the low rent element of gas exports. In some cases, rent on gas exports is almost nil at present c.i.f. prices of LPG and LNG. The costs of gathering, liquefaction, transportation, and distribution are such that the net returns are only sufficient to cover capital and operating requirements. This has led to a reexamination of a number of export projects in producing countries in favor of greater domestic utilization in energy-intensive industries, such as aluminum projects and petrochemicals, as well as in increased reliance on gas for household uses. In the course of this process the energy systems of OPEC nations may gradually shift from the use of petroleum products to gas in areas where such replacement is feasible. The net result of this trend might mean higher growth rates in the future for gas utilization and slightly lower growth rates for the consumption of petroleum products.

The income elasiticity of demand for energy in oil-exporting countries has consistently been higher than 1 and could possibly reach 2 in some cases. With accelerated development it is unlikely that the oil exporters will pay much attention to conservation. Saturation in energy consumption will only come about with economic maturity. It therefore can be expected that relatively high growth rates in energy consumption will persist for the whole of this century (some forecasts are needed).

The technique of discounting the future to optimize the present is of little value here. The purpose of this technique is to achieve economic development while drawing on the natural resource in order to have a viable economic base for self-sustaining growth without oil revenues at the time of that resource's depletion. Any discount rate appropriate for the period of imminent depletion would completely blur our vision of the early part of the next century when we want to see clearest.

Here we are concerned with a future point when production is declining to the extent that it is squeezing internal consumption after a period of fading exports. Such a point will come sooner or later for all OPEC countries. The magnitude of reserves does not prolong the life-span of production, if production levels are high. Continuous monitoring of output levels is needed to avoid pitfalls in the future resulting from too early depletion.

This "tail-end" period would be characterized by two main features: (1) a large segment of the economy would be using hydrocarbons as raw material; and (2) a great deal of capital stock would be geared to energy-intensive utilization. These two aspects make it imperative for planners to avoid early depletion of natural resources which would result in having to resort to petroleum imports at prices that would be crippling to the economy.

As greater diversification and economic maturity are attained, increasing attention should be paid to improving the efficiency of energy utilization. The nature of the choice of technology should gradually be changed, resulting in a more conservation-oriented capital stock and a more conservationist life-style. It also goes without saying that greater economic diversity should allow for nonpetroleum exports to compensate for the future decline of oil revenues. Such policies should not be too difficult to achieve if they are coupled with a more efficient use of oil revenues in the development process and with increasing oil prices in real terms.

As pointed out already, domestic energy needs are an increasing segment of total production. Initially, production was (and in most cases still is) growing faster than domestic consumption in oil-exporting countries. Domestic consumption will gradually encroach on oil available for exports. For OPEC as a whole this might be the case by the mid-1980s. For some nations (Indonesia, Venezuela, Algeria) this is already a reality. To maintain the level of exports by huge investments for enhanced recovery, high-cost small fields, or opening up new areas is self-defeating, unless it is accompanied by substantial price rises.

A view of declining net exports and possibly declining net revenues should not be clouded by the apparent vastness of reserves in some countries. Net revenues could decline not only if costs rise faster than prices, but also when domestic consumption eats into exports. Present production levels put many of the OPEC states into close time horizons, with most of them, except Kuwait, reaching maturity of production level now.

Domestic Energy Prices

Most oil-exporting countries are not market economies. They all exercise planning, and governments intervene extensively in most aspects of the economy. The public sector is already leading in all OPEC states, at least in industry. Therefore, prices do not play the same role in allocating resources as they do in "open" economies. But despite this, energy prices are significant as planning guidelines and on the exchange and distribution levels. They also influence disposable incomes through individual demand schedules.

Logically, prices should reflect opportunity cost. Any social cost should be identified, quantified, and accounted for; deliberate and conscious action should be taken by policy makers to offset it. This means that internal prices should equal f.o.b. export product prices. Subsidies should then be decided and charged to carefully worked out accounts.

Gasoline prices are already well above their f.o.b. price level. Due to its very low price elasticity, gasoline lends itself to being used as a means for collecting government revenue and for achieving income redistribution. The most difficult question arises with regard to fuel oil. It is argued that fuel oil prices should be kept low in order to increase the competitiveness of export industries or to encourage industrial investment in the private sector.

Another argument for low fuel oil prices is the necessity of controlling domestic inflation. Apart from the last argument, which implies that higher fuel oil prices would cause higher output prices on the market, there seems to be no reason for hiding a cost deliberately incurred by the economy. Internal prices can be protected and there are many ways to encourage domestic investment. Export subsidies have been used by many industrial countries in order to ensure their competitiveness. The inflation argument is not totally convincing, since inflationary pressure is better counteracted by direct price support, which allows better accounting and greater general control.

The planning of domestic petroleum prices is necessary for avoiding fluctuations; f.o.b. prices need not be followed automatically. In the case of gas it can be argued that its opportunity cost (shadow price) is zero or negative when it is flared. But even when flaring is replaced by exports, wellhead prices can be negligible. In the case of gas one might argue that wellhead pricing may be preferable to equalizing f.o.b. prices on the domestic market, since this way the domestic consumers would not be charged for the cost of export facilities for LPG and LNG.

Conclusions

The market for internationally traded crude oil is full of imperfections. Price formation has never been really determined by market forces. It is not reasonable not to rely on such forces to generate optimal prices for oil; prices in the past were a result of conflicting interests. Future prices of crude oil may still reflect the relative influence of such interests. In this situation it may be imperative that oil prices should be planned in advance.

Apart from the assumption that a mechanism is devised for adjusting the nominal price to maintain its value in real terms, price planning should take into account the fact that price elasticity of demand in the short term is very low. In the medium and long terms it may be higher for products which have some substitutes such as fuel oil for power stations. On the other hand, income elasticity of demand for energy is very high. The level of economic activity determines total demand for energy, and price levels—through their effect on economic activity—will influence demand. Prices also would affect demand through their impact on substitution and conservation.

In the short term, prices of crude oil could rise to reflect the levels of final products without significant effect on demand. They should be geared as well to meet the long-term objective of price planning in light of replacement and substitution on the one hand, and exhaustion on the other. In the long term, prices should not be geared only to the level of full substitutes (heavy oils and liquefied coal), but should be high enough to implement an orderly conversion from one type of fuel to another, reflecting the degree of depletion and scarcity.

For petroleum-exporting countries, oil prices are ultimately measured by the degree of economic development achieved through the transformation of oil-generated revenues to fixed capital formation and social development. In this connection, production levels and prices are closely tied to the development process. Depletion should not be allowed to set in before the economy is capable of self-sustained growth. Revenues are ideally maximized through the unit price, not through aggregate exports. The interaction among price, development, and depletion (production) is the pivot to the policies of exporting countries. A direct corollary to the price level as viewed by the producing nations is the burden of investment for further development of the oil sector which competes with general economic development for revenues. This may be resolved by gearing prices in the medium term to marginal costs of replacement (supply cost).

Energy-intensive choices of technique coupled with very low product prices in OPEC countries are leading to very high growth rates in energy consumption. As domestic consumption builds up in the medium and long run, domestic demand might begin to compete with exports for total production. Sufficient attention should be paid to these problems in deciding on present production policies. Additionally, sufficient attention should be given to the pricing of petroleum products and gas at home. Ideally, such prices should reflect the f.o.b. (or wellhead export value of such products. Any subsidies may be decided upon with full awareness by the policy makers to the real cost of such subsidies.

Appendix

Table 1

AVERAGE PRICES OF A REFINED BARREL OF OIL IN
EUROPEAN COUNTRIES, 1962-1978
(in U.S. $/barrel)

	Retail		Wholesale	
Year	Price	Tax	Price	Tax
1962	12.03	6.77	11.88	6.77
1963	12.07	6.03	11.82	6.03
1964	11.90	6.03	11.64	6.03
1965	12.00	6.49	11.74	6.49
1966	12.18	6.79	11.91	6.79
1967	12.66	6.97	12.38	6.97
1968	12.64	7.00	12.35	6.99
1969	12.96	7.59	12.63	7.58
1970	12.66	6.91	12.35	6.90
1971	14.55	7.36	14.19	7.34
1972	15.73	8.13	15.30	8.11
1973	18.98	9.22	18.50	9.19
1974	26.62	11.05	26.17	11.04
1975	30.97	13.30	30.35	13.28
1976	32.27	13.69	31.54	13.66
1977	36.05	14.79	35.73	14.79
1978	40.69	16.57	40.37	16.56

Table 2

EFFECT OF EXCHANGE RATES, INFLATION, AND THEIR COMBINED EFFECT ON MARKER CRUDE PRICES, 1974-1978
(December 1973=100)

		Exchange Rate	Inflation	Combined
1974	Jan	93.4754	101.36	94.75
	Feb	97.5628	103.29	100.77
	Mar	100.0267	106.70	106.73
	Apr	101.3403	109.89	111.36
	May	102.8117	113.19	116.37
	Jun	101.3664	115.85	117.43
	Jul	101.3360	117.93	119.51
	Aug	100.1573	118.95	119.14
	Sep	99.2656	122.40	121.50
	Oct	99.4261	124.79	124.07
	Nov	100.7722	128.47	129.46
	Dec	102.5119	128.10	131.32
1975	Jan	104.8747	128.48	134.74
	Feb	106.3642	130.97	139.31
	Mar	107.4245	132.59	142.43
	Apr	105.7592	133.35	141.03
	May	105.9931	133.47	141.47
	Jun	106.4491	132.33	140.86
	Jul	102.7096	133.73	137.35
	Aug	99.8559	134.25	134.06
	Sep	98.6226	133.84	132.00
	Oct	98.9735	134.01	132.63
	Nov	99.1373	134.08	132.92
	Dec	98.4252	133.31	131.21
1976	Jan	98.6948	133.40	131.66
	Feb	98.5992	133.52	131.65
	Mar	97.4602	134.15	130.74
	Apr	96.9587	134.15	130.07
	May	96.8638	134.15	129.94
	Jun	96.4899	136.05	131.27
	Jul	96.6120	137.61	132.95
	Aug	96.9884	138.87	134.69
	Sep	97.6445	139.75	136.46
	Oct	98.0338	140.06	137.31
	Nov	97.9351	140.79	137.88
	Dec	97.2956	142.80	138.94

Table 2

EFFECT OF EXCHANGE RATES, INFLATION, AND THEIR COMBINED EFFECT ON MARKER CRUDE PRICES (continued)
(December 1973=100)

		Exchange Rate	Inflation	Combined
1977	Jan	97.3282	144.61	140.75
	Feb	97.0303	146.32	141.97
	Mar	97.1172	149.27	144.97
	Apr	97.4319	150.48	146.62
	May	97.5855	151.24	147.59
	Jun	97.6412	151.23	147.66
	Jul	99.2036	151.75	150.54
	Aug	98.6138	152.61	150.49
	Sep	97.8210	153.45	150.11
	Oct	99.3277	153.60	152.57
	Nov	100.6014	154.22	155.15
	Dec	102.9972	152.92	157.50
1978	Jan	104.8835	152.05	159.48
	Feb	105.9782	150.57	159.57
	Mar	107.1846	151.88	162.79
	Apr	107.3989	152.74	164.04
	May	105.4307	153.48	161.81
	Jun	107.1668	153.24	164.22
	Jul	109.6980	154.44	169.42
	Aug	113.1010	156.94	177.50
	Sep	113.7378	156.55	178.06
	Oct	118.0658	154.69	182.64
	Nov	114.3848	155.39	177.74
	Dec	114.1943	156.11	178.27

Chart 1

SEPARATING THE EFFECTS OF INFLATION AND EXCHANGE RATES
ON THE MARKER CRUDE PRICE

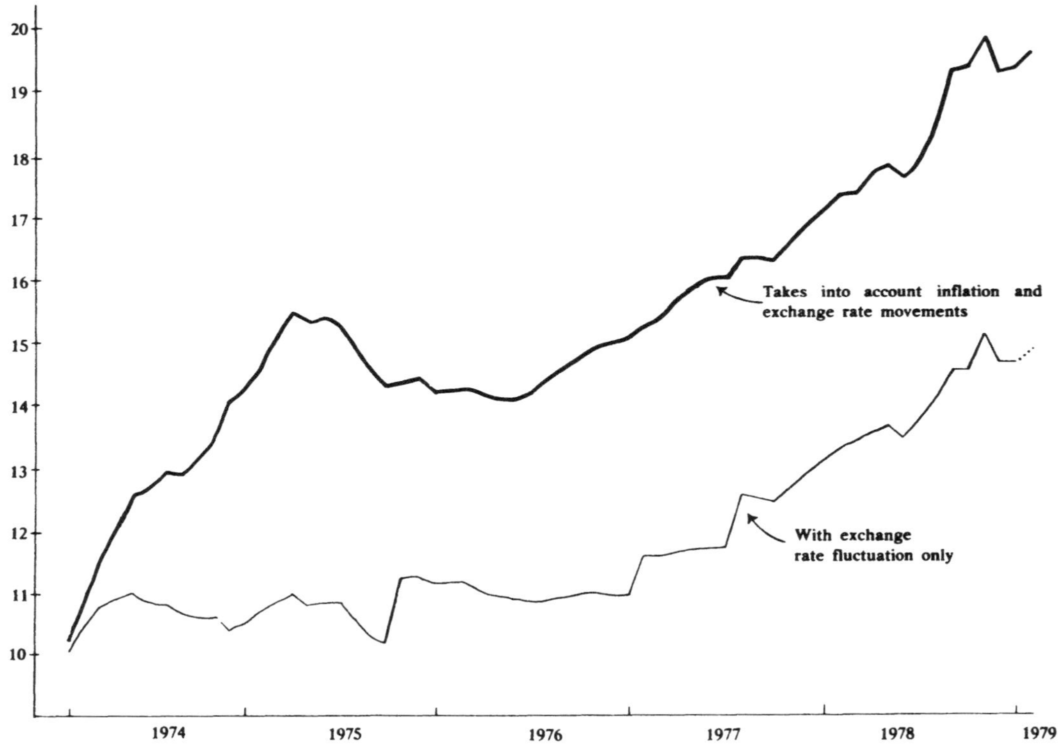

Table 3

OPEC IMPORT PRICE INDEX, 1974-1978
(by quarters)

		Base: 1973=100	Base=Previous Quarter
1974		135.84	
1975	I	153.98	113.35
	II	163.00	105.86
	III	168.88	103.61
	IV	177.49	105.10
1976	I	186.69	105.18
	II	191.51	102.58
	III	200.65	104.77
	IV	211.30	105.31
1977	I	223.19	105.63
	II	237.19	106.27
	III	251.89	106.20
	IV	268.79	106.71
1978	I	283.60	105.51
Forecasts			
1978	II	307.51	
	III	325.93	
	IV	345.46	

Table 4

NOMINAL, ACTUAL, AND DEFLATED MARKER CRUDE PRICES, 1975-1978
(by quarters, in U.S. $/barrel)

	I	II	III	IV
1975				
Nominal	16.68	17.66	18.30	19.23
MCP	10.46	10.46	10.46	11.51
Deflated	6.80	6.42	6.20	6.48
1976				
Nominal	20.23	20.75	21.74	22.89
MCP	11.51	11.51	11.51	11.51
Deflated	6.17	6.01	5.74	5.45
1977				
Nominal	24.18	25.70	27.29	29.12
MCP	12.09	12.09	12.70	12.70
Deflated	5.42	5.10	5.04	4.72
1978				
Nominal	30.73	33.32	35.31	37.43
MCP	12.70	12.70	12.70	12.70
Deflated	4.48	4.13	3.90	3.68

Chart 2

NOMINAL, ACTUAL, AND DEFLATED MARKER CRUDE PRICES
(based on the OPEC import price index)

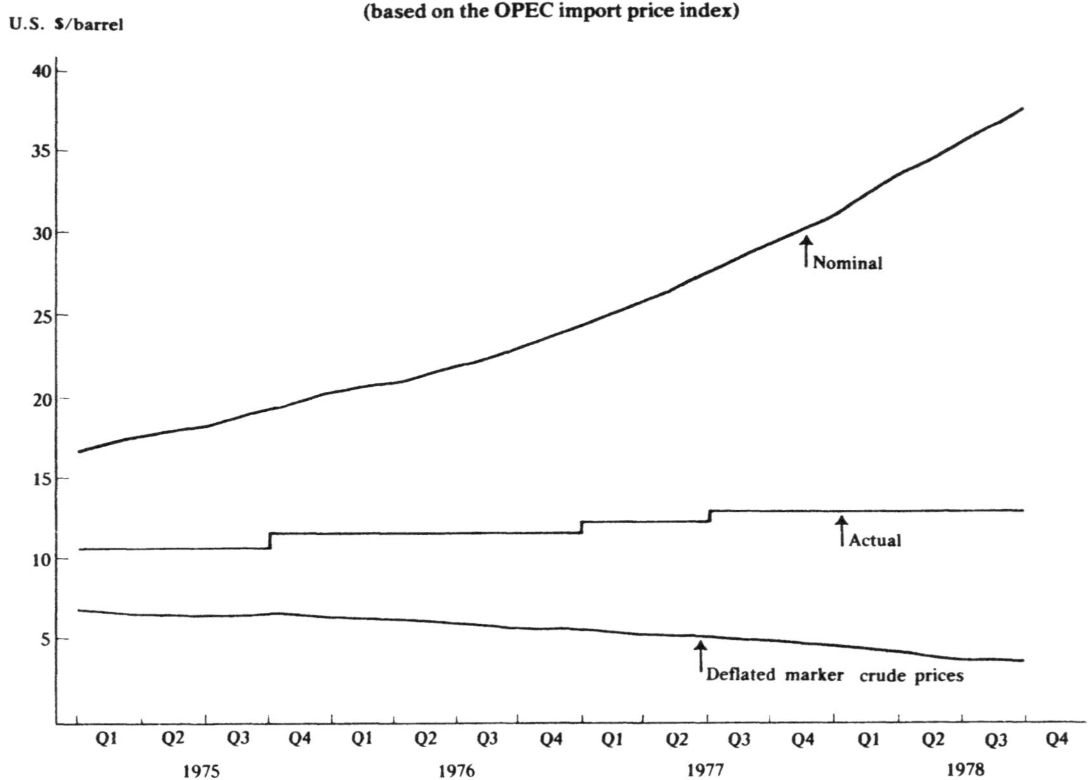

Chart 3

NOMINAL, ACTUAL, AND DEFLATED MARKER CRUDE PRICES[a]

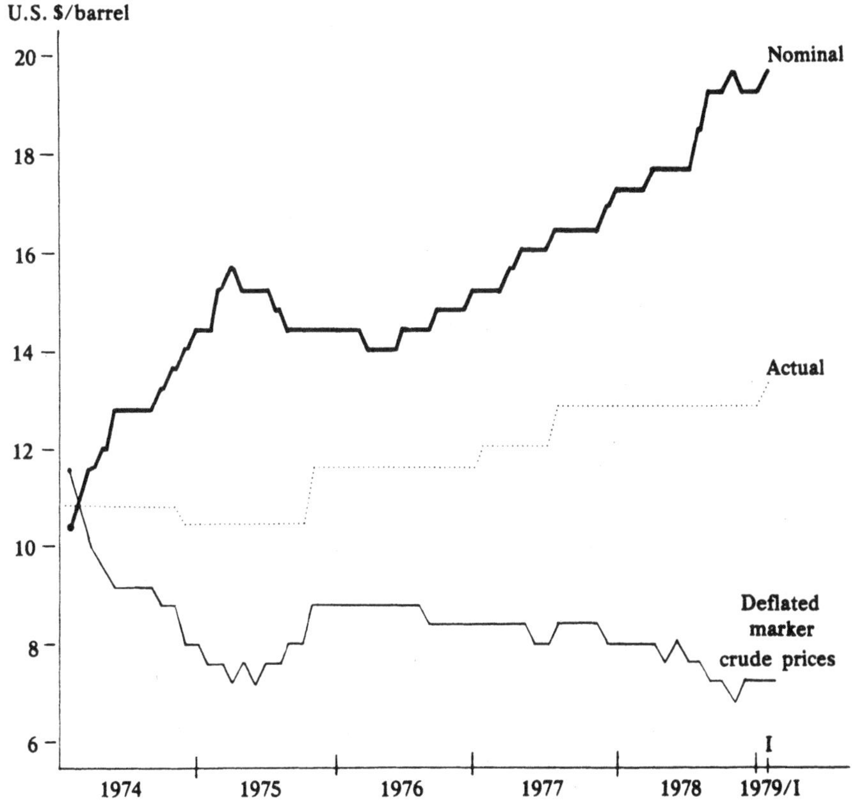

[a]Based on the combined effects of export price indices of industrialized countries and exchange rate fluctuations of the United States dollar, i.e., using the 11 currencies of the Geneva II basket plus the U.S. dollar, weighted in accordance with OPEC imports from the countries concerned.

Table 5

OPEC OIL PRODUCTION[a] NECESSARY TO COVER GROSS FIXED CAPITAL FORMATION AND DOMESTIC CONSUMPTION

Year	Gross Fixed Capital Formation (million U.S. $)	Exports Necessary to Cover Gross Fixed Capital Formation (1,000 b/d)	Domestic Consumption (1,000 b/d)	Production Necessary to Cover GFCF and Domestic Consumption (1,000 b/d)	Actual Production (1,000 b/d)	Production Necessary to Cover GFCF and Domestic Consumption As a Percentage of Actual Production
1977	97.232.3	24,701.7	1,789.6	26,491.3	28,955.0	91.5
1976	79,197.5	21.384.1	1,529.7	22,913.8	28,304.0	81.0
1975	59,731.8	16,803.8	1,659.1	18,462.9	26,315.5	70.2
1974	35,750.7	14,377.0	1,440.8	15,817.8	28,532.2	55.4
1973	24,040.8	29,680.0	1,245.2	30,925.2	28,885.6	107.1
1972[b]	17,526.1	32.076.7	931.1	33,007.8	25,331.2	130.3
1971	13,303.0	28,804.4	1,125.6	29,930.0	23,852.4	125.6
1970	10,927.6	35,851.9	904.3	36,756.2	22,266.9	165.1
1969	9,991.1	34,904.4	784.7	35,689.1	19,918.6	179.2
1968	8,458.2	26,884.4	819.8	27,704.2	17,944.5	154.4
1967	7,001.7	24,926.3	746.5	25,672.8	16.137.9	159.1
1966	6,104.7	26,070.8	625.8	26,696.6	15,111.6	176.7
1965	5,439.5	23,899.1	636.3	24,535.4	13,816.6	177.6
1964[c]	4,623.7	21,359.5	538.9	21,898.4	12,574.2	174.2
1963	4,382.5	20,509.2	416.6	20,925.8	10,823.7	193.3
1962	4,095.5	19,873.0	401.0	20,274.0	9,847.3	205.9
1961[d]	3,931.8	21,327.1	473.4	21,700.5	8,754.3	247.9
1960	3,894.6	21,595.1	393.4	21,988.5	8,105.2	271.3

[a]Excluding Qatar and United Arab Emirates.
[b]Ecuador included as of 1972.
[c]Indonesia included as of 1964
[d]S.P. Libyan A.J. included as of 1961.

Chart 4

TOTAL OPEC: CRUDE OIL PRODUCTION VS. PRODUCTION NECESSARY TO COVER GROSS FIXED CAPITAL FORMATION AND DOMESTIC CONSUMPTION, 1960-1977

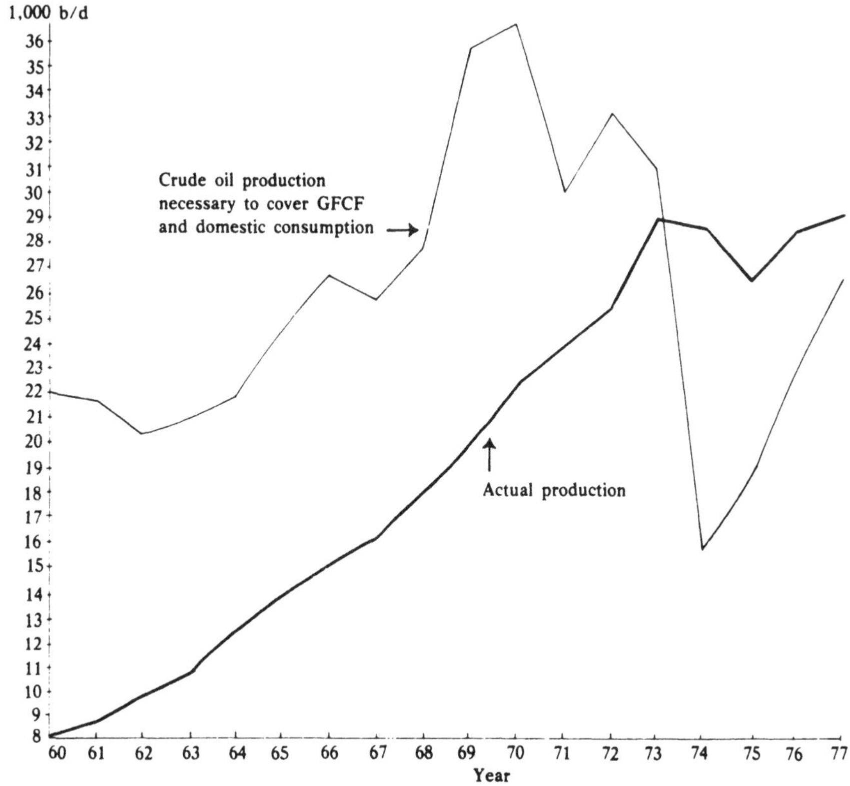

Table 6

OPEC OIL PRODUCTION NECESSARY TO COVER DEVELOPMENT EXPENDITURE, 1970-1977

Year	Development Expenditure (million U.S. $)	Exports Necessary to Cover Development Expenditure (1,000 b/d)	Domestic Consumption (1,000 b/d)	Production Necessary to Cover Development Expenditure and Domestic Consumption (1,000 b/d)	Actual Production (1,000 b/d)	Production Necessary to Cover Development Expenditure As a Percentage of Actual Production
1977	66,690.0	15,871.1	1,839.8	17,710.9	31,398.3	56.4
1976	53,909.5	14,287.0	1,574.4	15,861.4	30,737.7	51.6
1975	43,690.0	12,256.0	1,697.5	13,953.5	27,192.7	51.3
1974	26,784.9	9,729.7	1,457.4	11,187.1	30,729.2	36.4
1973	11,614.9	15,330.1	1,263.1	16,593.2	30,988.5	53.5
1972[a]	8,138.7	14,861.0	936.8	15,797.8	27,094.4	58.3
1971	5,926.0	12,692.9	1,134.2	13,827.1	25,322.6	54.6
1970	4,602.3	14,622.5	908.1	15,530.6	23,408.9	66.3

[a]Ecuador included as of 1972.

Chart 5

TOTAL OPEC: CRUDE OIL PRODUCTION VS. PRODUCTION NECESSARY TO COVER DEVELOPMENT EXPENDITURE AND DOMESTIC CONSUMPTION, 1970-1977

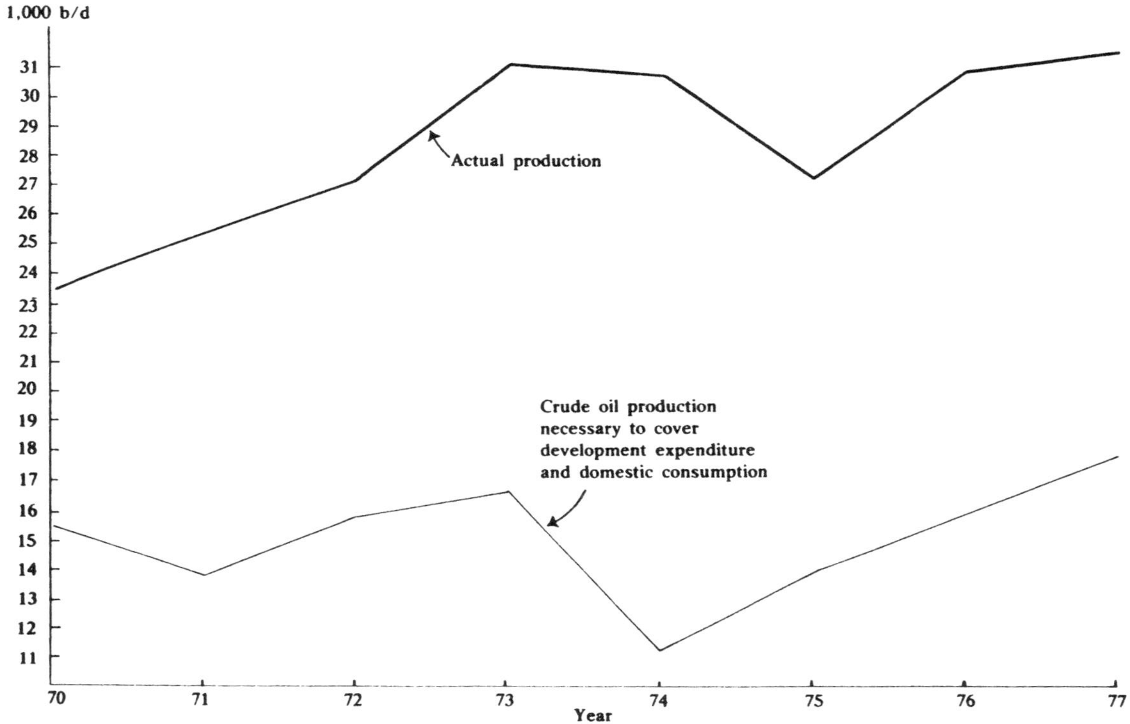

Table 7

AVERAGE PRICES OF A REFINED BARREL OF OIL, WEIGHTED OPEC TOTAL
(in U.S. $/barrel)

	Average Price	Average Tax	% of Total OPEC Demand
Motor gasoline:			
Premium	22.93	3.07	13.1
Regular	15.54	4.14	18.3
Household kerosene	6.17	0.04	17.8
Distillate fuel oil	7.42	1.73	30.2
Residual fuel oil	3.26	0.10	20.6
Total average	9.87	1.71	100.0
Average price excl. tax	8.16		

5

AN ECONOMIC ANALYSIS OF CRUDE OIL PRICE BEHAVIOR IN THE 1970s

*Walter J. Mead**

The behavior of crude oil prices in the 1970s presents a record of apparent chaos. The roughly fourfold increase in the price of crude oil from 1971 through 1974 was partly responsible for nearly worldwide inflation and recessions in 1974 and 1975. Because oil is a major primary energy source, higher world crude oil prices led to similar price increases in substitute energy forms. The prices of energy-intensive products in turn reflected these cost increases. Consumers, unaware of rapid increases in the money supply in the United States and most other leading nations, tended to blame the crude oil price increase for general inflation.

In the 1970s, economists have passed through three phases in their thinking about the optimum price of oil. First, in the early part of this decade, static micro-theory was frequently applied, indicating that the competitive price should equal the marginal cost for marginal supplies. Since Middle Eastern producers appeared to have extremely large crude oil reserves, having marginal costs of less than $1 per barrel, some observers asserted that Middle Eastern prices should be limited to their marginal cost and that in the long run competition would produce this result.

Second, realizing that oil reserves were limited and that profit-maximizing producer nations would foresee exhaustion, some economists applied capital theory and computed the optimum present price on the basis of assumed backstop technologies and their probable future prices.[1]

---

*Dr. Walter Mead is Professor of Economics at the University of California, Santa Barbara. He served as President of the Western Economic Association from 1958 to 1959 and has carried out research over the years in industrial organization and natural resources with a decade or more in the energy field. Professor Mead was Senior Economist with the Ford Foundation Energy Policy Project. His publications have appeared in the *American Economic Review*, *Land Economics*, *Natural Resources Journal*, and *Journal of Energy and Development*, among others.

[1]For example, see William D. Nordhaus, "The Allocation of Energy Resources," *Brookings Papers on Economic Activity*, no. 3, 1973, pp. 529-76.

Third, a missing link was provided by Johany who combined capital theory and property rights theory to explain the sharp increase in crude oil prices that occurred when oil reserves were nationalized in 1973.[2]

The purpose of this paper is to review the record of crude oil pricing in the light of capital theory and property rights theory, including the record of oil price behavior resulting from the Iranian supply reductions occurring in the winter of 1978-1979. The analysis indicates that the Iranian "crisis" should have no long-run effect on crude oil prices beyond the expected steady upward trend in the real price of crude oil.

The Record of Crude Oil Price Behavior

In the two decades from 1950 to 1970, large new discoveries of crude oil occurred primarily in the Middle East. These new discoveries involving low marginal costs created the impression of immense oil reserves and resources available to the world. Production from these new reserves maintained downward pressure on prices such that the real price of crude oil declined moderately as shown in table 1. The oil industry in the United States successfully appealed to government for relief from competition. The government response took the form of tighter restrictions on output through market-demand prorationing[3] and restrictions on foreign supplies through the import quota system.

With the Arab-Israeli war of 1973 and its consequent oil embargo, together with widespread nationalization of crude oil production, the average wellhead price of crude oil in the United States increased from $3.39 per barrel in 1971 to an imported (uncontrolled) price of $13.93 in 1975. The embargo reflected the temporary unity of Arab members of OPEC only or more properly of OAPEC (Organization of the Arab Petroleum Exporting Countries). Non-Arab members of OPEC expanded oil production during the period of the embargo. With the end of the war and the embargo, responsibility for coordinating output and determining price still rested with OPEC. The conventional wisdom holds that OPEC is a cartel seeking to achieve monopoly profits for its members. The record of prices from the end of the embargo through June 1978 shows an upward trend in nominal prices for imported oil in the United States, but a declining trend in the real price of oil. This weakness in crude oil prices relative to wholesale prices of all commodities reflects what may be a "shake out"

---

[2]Ali D. Johany, "OPEC Is Not a Cartel: A Property Rights Explanation of the Rise in Crude Oil Prices" (Ph.D. dissertation, University of California, Santa Barbara, 1978).

[3]The legislative basis for market-demand prorationing was laid in the 1930s.

phase for the OPEC "cartel" and raises doubts about the real impact of the cartel on oil prices in mid-1978. By mid-1978 increasing oil production in response to higher prices put downward pressure on world prices on crude oil. Widespread evidence appeared of minor price concessions as well as nonprice concessions in the form of easy credit terms.

Table 1

UNCONTROLLED CRUDE OIL PRICES IN THE UNITED STATES
(in $/barrel)

	Average U.S. Wellhead Price	Wholesale Price Index (1967=100)	Real Price U.S. Wellhead (1967 dollars)
1950	2.51	81.8	3.07
1960	2.88	94.9	3.03
1970	3.18	110.4	2.88
1971	3.39	113.9	2.97
Imported oil ("Refinery acquisition cost")			
1975	13.93	174.9	7.96
1976	13.48	183.0	7.37
1977	14.53	194.2	7.48
June 1978	14.54	209.6	6.94
Jan-March 1979	15.00 - 22.00[a]	222.0[a]	6.76 - 9.91

Sources: 1950-1971, United States Bureau of Mines; 1975-1978, United States Department of Energy, *Monthly Energy Review*.

[a]Estimate.

But any downward pressure on price ended in October 1978 with the beginning of political problems in Iran, resulting in progressively lower rates of Iranian oil production and shipments. The Iranian crisis led to anticipated shortages and consequent price increases. Price increases took the form of surcharges and sharply higher spot prices for crude oil not covered by binding long-term contracts. These spot prices appeared to have reached $22.00 per barrel for isolated sales in early 1979.

Capital Theory Applied to Crude Oil Pricing

Under competitive conditions, owners of nonrenewable crude oil reserves are expected to maximize the present value of their future expected flows of net income from oil production. This present value maximization process under equilibrium conditions would produce an upward trend in the *in situ* value of crude oil reserves equal to the opportunity cost of money. This orderly trend would be periodically shifted as new information becomes available concerning such factors as new discoveries, new low-cost energy technologies, shifts in demand, and the like.

Applying capital theory requires use of a discount rate equal to the opportunity cost of capital. In the United States the real opportunity cost of capital appears to be in the range 2 to 3 percent. At 8 percent inflation, this real interest rate corresponds with a nominal interest rate of 10 to 11 percent. For the major oil-producing countries of the Middle East, including Saudi Arabia, Kuwait, and the United Arab Emirates, the real opportunity cost of capital would appear to be no more than 2 percent. Investment opportunities for large capital flows in these countries are not attractive and diminishing returns quickly limit domestic investment opportunities. The best alternative use of capital for these major oil-exporting nations would appear to be foreign investments mainly in the United States and Western Europe. But the rate of return on such investments is no better than that earned by nationals in those countries and is likely to be even lower due to exchange rate depreciation.

Figure 1 illustrates the application of capital theory under conditions of secure property rights (this latter issue will be discussed below). At a 2 percent real return on capital, a price of $14.50 per barrel must increase to $39.00 per barrel in 50 years. Under conditions of the Iranian crisis, a current price of $17.00 per barrel must rise to $45.76 per barrel in 50 years in order to yield a 2 percent real return to the resource owner.

If one believes that the deflated price of large supplies of substitute fuel in 50 years is likely to be about $40.00 per barrel, then the rational present value of these reserves discounted at 2 percent real discount rate is $14.86 per barrel. A $17.00 per barrel Iranian crisis price is unsustainable under conditions of a $40.00 price for a backstop technology in 50 years because the implied rate of return for holding reserves is only 1.7 percent.

If oil-producing countries are profit maximizers and adopt the assumptions given above, then they would conclude that it is more profitable to increase present production, convert their oil into cash, and invest their funds in their best alternative uses yielding 2 percent. The result of this action by producers is also shown in figure 1. Present world production would be expanded causing present prices to decline relative to future prices until the expected growth in the net value of oil resources equalled the opportunity cost of capital. This adjustment process would come to an end when present prices are in the range $14.50 - $15.00, given the assumed $40 per barrel backstop technology price 50 years later.

Shale Oil As a Price Limiting Energy Source

While shale resources are huge and are concentrated in the United States, Russia, and China, the history of oil shale places it in a group of

Figure 1

OPTIMUM PRICE TRENDS OVER TIME

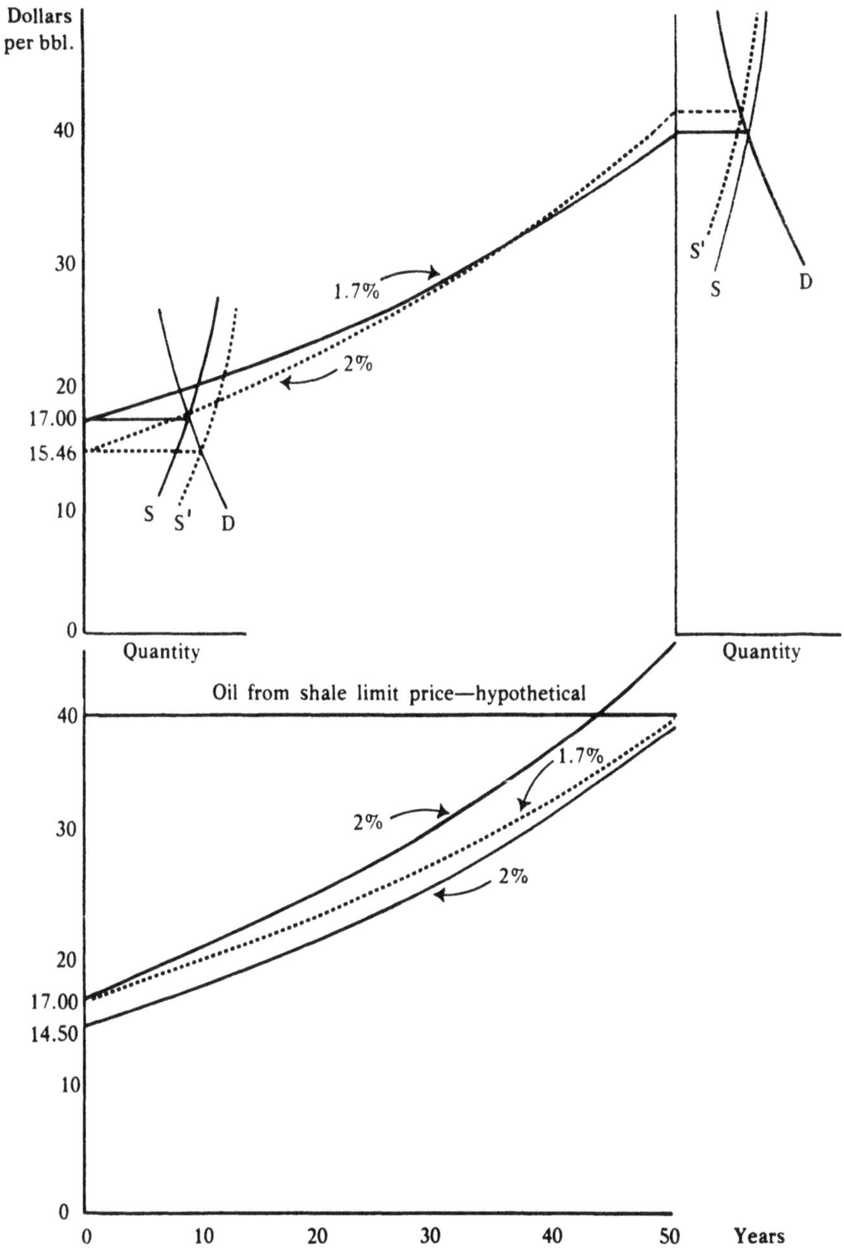

innovations perpetually on the threshold of production. Under pressure from environmental restrictions, water supply problems, uncertain costs, and escalating capital costs for plant construction, costs of producing shale oil have increased in tandem with and somewhat ahead of crude oil prices. Current rough estimates indicate that production of oil from shale resources in the United States would be feasible with oil prices in the range of $18 to $22 per barrel.

Given the enormous size of worldwide oil shale reserves, once production from shale becomes profitable, its real price is likely to remain constant and the appropriate pricing theory would be static micro-theory. Under conditions of rapid technological advances associated with application of a new technology, together with conventional "learning curve" cost behavior, the real price of producing oil from shale should be expected to decline in the initial two or three decades of production.

Crude oil producing nations interested in maximizing the present value of their crude oil resources must consider oil shale as a price-limiting energy source. If one assumes that the real cost of producing oil from shale will eventually stabilize at $40.00 per barrel and that the supply of crude oil in the world for which oil shale is a substitute will be exhausted in approximately 50 years, then at a 2 percent real discount rate the optimal *in situ* price of crude oil in the present is $14.86. Alternatively, if one believes that the real future cost of oil from shale will be $30.00, then under the same conditions identified above the optimal present value of crude oil *in situ* is $11.15 per barrel. Thus, given these future prices for shale oil, there is not much room for further real crude oil price increases on a sustainable basis.

In addition, crude oil producing nations may consider other backstop technologies. These include coal liquefaction, nuclear fission, and fusion. Of these three sources, nuclear fission alone is currently feasible and operating. Large-scale power production from fission is restrained by fears of accidents, nuclear proliferation, and high level waste storage safety. The March 1979 accident at the Three Mile Island nuclear plant in Pennsylvania will contribute both to further delays in nuclear development and to higher costs for this backstop technology. To the extent that fission power is the relevant backstop technology, higher costs for redundant safety measures will have the effect of raising the rational price of crude oil today.

A Capital Theory and Property Rights Theory Interpretation of the Fourfold Increase in Crude Oil Prices

The words "OPEC cartel," like "damn Yankee," tend to be a single inseparable expression. There are many economic groups interested in promoting the concept of an OPEC "cartel." Member states believe that OPEC has cartel power, otherwise occasional threats to increase oil prices unless certain conditions are met would be meaningless. In the absence of the conviction that a cartel exists there would be a loss of bargaining power in negotiations concerning such matters as military equipment and Middle Eastern politics. Without cartel power, OPEC meetings to discuss crude oil prices would be a charade. The OPEC bureaucracy is more secure if the world believes in the cartel concept. Government officials, especially in the United States, find the cartel concept a convenient rationalization for continued price controls. Part of the argument in favor of crude oil price controls rests on the point that without controls the OPEC cartel would set United States energy prices.

In order to evaluate the pricing power of OPEC we need to specify the technical economic definition of a cartel and to identify how a cartel is expected to operate. From a technical economic theory point of view, a cartel is a collusive combination of a small number of producers or sellers of a homogenous product acting in harmony in order to achieve some price or other economic objective. An effective cartel is able to collusively reduce output in order to raise selling price above a level that would be attained by competition alone. Output control is the essential ingredient for effective cartel (monopoly) power. Historically the United States oil industry taught the world how to control oil price through control over output. In the mid-1930s the American oil industry, with strong support from their politicians, persuaded the United States Congress to enact legislation providing for the Interstate Oil Compact Commission. This legislation, in turn, authorized individual oil-producing states to regulate oil produciton within state borders in the name of conservation.

Without police powers, effective regulation of output is impossible, except in a highly concentrated industry. Accordingly the oil industry again persuaded Congress in 1935 to enact the Conally Hot Oil Act. This important legislation provided that any producer not in compliance with state prorationing regulations would not be permitted to sell oil in interstate commerce. Thus the power of the federal government was made available to enforce state market-demand prorationing regulations. The power of government was used to enforce output restrictions with the result that prices were raised above competitive levels.

Where is the power within OPEC to allocate production cutbacks among member states? What is the enforcement mechanism? Where are the courts? Where is the OPEC army? And what is its fire power? A former advisor to a Prime Minister of Iran reported that the United States style market-demand prorationing had indeed been tried in OPEC and failed.[4]

[4] Statement by Dr. Fereidon Fesharaki at the Fifth International Conference on Energy in the 1980s, October 16, 1978, University of Colorado, Boulder. The proceedings of that conference will be published as *Energy in the 1980s: Conflict or Cooperation?* (Boulder, Colorado: International Research Center for Energy and Economic Development, forthcoming 1979).

Without an effective mechanism for coordinated control over output by separate producers, there is no theoretical basis for believing that cartel pricing is workable.

This theoretical conclusion should be tested against the facts. Economic theory of collusion would lead one to expect that if collusion is effective the members of the cartel would reduce output during periods of market weakness in order to obtain their price objectives. In order to test this expectation we may use the period from 1974 through 1977. The base period in this case follows the five-month Arab oil embargo which started on October 15, 1973. The base year 1974 was a year of strong markets, clearly qualifying as a seller's market. A worldwide depression started in late 1974 and extended through 1975. But 1976 and 1977 continued to be years of market weakness characterized by widespread price cutting or nonprice concessions by individual OPEC members. Effective collusion would require output restraint and relatively stable market shares.

The record of individual OPEC member country output is shown in table 2. We find that six OPEC members producing approximately one-half of total OPEC output expanded production such that their market shares increased from 52.2 percent to 58.8 percent of OPEC production. Another seven members of OPEC reduced output such that their market shares declined from 47.8 percent to 41.2 percent of total OPEC production. This pattern is not the kind of record one would expect to find under an effectively collusive cartel. However, the record is mixed. All OPEC members together increased their output by 2 percent. Non-OPEC members increased their output by 12 percent. Therefore, the OPEC market share in total declined from 54.6 percent to 52.2 percent of world crude oil production. This later evidence is consistent with cartel behavior.

Table 3 looks at the same data in another way. Robert Pindyck[5] and others have classified OPEC members into "saver countries" and "spender countries." One would expect that the saver countries might be willing to bear the burden of output reduction, permitting spender countries who could not afford output reductions to maintain or increase their level of output. The record, however, shows that saver countries in total expanded output and market shares at the expense of the spender countries.

Theodore Moran has classified OPEC member countries into a "core of balancer" countries and "competitive fringe" countries.[6] The former group consists of countries that "adjust production to meet the remaining

[5]Robert S. Pindyck, "OPEC's Threat to the West," *Foreign Policy*, spring 1978, pp. 36-52.

[6]Theodore Moran, *Oil Prices and the Future of OPEC* (Baltimore: The Johns Hopkins University Press for Resources for the Future, Inc., 1978).

demand," while the latter group consists of "high-population, high-mobilization" countries that "prefer to produce oil and natural gas at a rate as close to full capacity as possible." This classification of OPEC countries is shown in table 4. In the period of market weakness from 1974 through 1977, the "core of balancer countries" in fact expanded output and market share, at the expense of all other countries.

Table 2

OPEC COUNTRY OUTPUT AND MARKET SHARES, CLASSIFIED BY EXPANDING AND CONTRACTING COUNTRIES

	1974		1977	
OPEC Member Country	Output Per Day (000 bbls)	Market Share in OPEC (percent)	Output Per Day (000 bbls)	Market Share in OPEC (percent)
Countries expanding output:				
Saudi Arabia	8,481	27.6	9,200	29.5
Iraq	1,975	6.4	2,265	7.3
Libya	1,521	4.9	2,080	6.7
Indonesia	1,375	4.5	1,685	5.4
United Arab Emirates	1,689	5.5	2,009	6.4
Algeria	1,009	3.3	1,123	3.6
Total	16,050	52.2	18,362	58.8
Countries contracting output:				
Kuwait	2,547	8.3	1,969	6.3
Iran	6,022	19.6	5,699	18.3
Venezuela	2,976	9.7	2,238	7.2
Nigeria	2,256	7.3	2,097	6.7
Others (3)	895	2.9	849	2.7
Total	14,696	47.8	12,852	41.2
Total OPEC	30,746	100.0	31,215	100.0
Total world	56,268		59,798	
OPEC share of world output		54.6		52.2

Source: American Petroleum Institute, *Basic Petroleum Data Book.*

The foregoing analysis of the record does not correspond with expectations based on standard models of collusive behavior. There is no evidence of coordinated control over output with constant market shares.

An alternative model, the dominant-firm price-leadership model, may also be tested over the same period. The dominant "firm" is generally agreed to be Saudi Arabia. Under the rules of this model the dominant producer would determine the price that is optimal from its point of view and permit all other producers to sell all they wish at the price which the dominant producer selects. Under assumptions of a highly inelastic total demand for crude oil and relatively flat marginal costs for the dominant firm, the logic of this model indicates that in a weak market the dominant firm would substantially reduce output and *raise* prices, while all other producers would expand output. However, reference again to table 2 shows that Saudi Arabia increased output and market share, both within OPEC and in total world markets.

Table 3

OPEC COUNTRY OUTPUT AND MARKET SHARES, CLASSIFIED BY "SAVER COUNTRIES" AND "SPENDER COUNTRIES"

	1974		1977	
	Output Per Day (000 bbls)	Market Share in OPEC (percent)	Output Per Day (000 bbls)	Market Share in OPEC (percent)
"Saver countries"				
Saudi Arabia	8,481	27.6	9,200	29.5
Kuwait	2,547	8.3	1,969	6.3
Libya	1,521	4.9	2,080	6.7
United Arab Emirates	1,689	5.5	2,009	6.4
Total	14,238	46.3	15,258	48.9
"Spender countries"				
Iran	6,022	19.6	5,699	18.3
Venezuela	2,976	9.7	2,238	7.2
Algeria	1,009	3.3	1,123	3.6
Indonesia	1,375	4.5	1,685	5.4
Nigeria	2,256	7.3	2,097	6.7
Ecuador	174	0.6	183	0.6
Total	13,812	44.9	13,025	41.7
Unclassified				
Iraq	1,975	6.4	2,265	7.3
Gabon	202	0.7	222	0.7
Qatar	519	1.7	445	1.4
Total	2,696	8.8	2,932	9.4
Total OPEC	30,746	100.0	31,215	100.0

Source: American Petroleum Institute, *Basic Petroleum Data Book.*

The evidence examined here does not support conventional collusive pricing models. On the other hand, given a relatively inelastic total demand for crude oil, it is obvious that Saudi Arabia has the power to influence price through variations in its output.

Instead of a cartel theory, there appears to be a better explanation of price and output behavior in the crude oil market. Nordhaus, Pindyck and others have applied capital theory to the crude oil market and have shown that prices prevailing at different points in time correspond closely with prices which would prevail under assumed competitive conditions.[7]

Apart from the short-term impact of the Arab (OAPEC)embargo, why did the price of oil increase from about $3.25 in 1973 to $14.50 in 1977 (landed in the United States)? The answer to this question requires a joining of two relevant principles — capital theory and property rights theory. A recent dissertation by Ali Johany[8] has shed new light on rational price behavior for crude oil prices assuming competitive conditions.

Johany observed that during the 1950s and 1960s there was a progressive awareness on the part of international oil companies holding oil concessions in the Middle East that their property rights were in jeopardy. Nationalization, or its euphemism "participation," was the apparent wave of the future. But as the concessionaire companies became increasingly fearful in their property rights, they naturally reacted by raising their discount rates. This means that production was shifted from the future to the present. The higher the discount rate, the greater will be the shift toward present production. This real fear of loss of property rights resulted in rapid increases in oil production from Middle Eastern sources. From 1960 through 1970 the compound annual growth rate in oil production from the Middle East was 10.9 percent. From 1970 through 1973, the compound annual growth rate was 15.0 percent. This is rational economic behavior for firms convinced that they faced imminent loss of property rights.

Reflecting these output increases, crude oil prices (real) declined 6 percent in 20 years from 1950 through 1970. Output increases were matched by worldwide growth in demand with only modest increases in nominal prices.

By the end of 1973, a total shift in property rights in fact had occurred. Host countries were in complete control of output by year-end 1973, and they were either unilaterally determining the terms of sale for their crude or nationalization had been completed.

But with property rights shifted from companies to host countries, security of property rights was reestablished. The relevant discount rate became that of the host country rather than the concessionaire company. Discount rates declined not only because property rights became secure in

[7]William D. Nordhaus, op. cit. and Robert F. Pindyck, "Gains to Producers From the Cartelization of Exhaustible Resources," *Review of Economics and Statistics*, May 1978, pp. 238-51.

[8]Ali D. Johany, op. cit.

the hands of the host country, but also because the opportunity cost of money for the host country was relatively low. Investment opportunities for large sums of money in Saudi Arabia are relatively unattractive. Investments abroad yield only competitive rates of return, but were further jeopardized by exchange rate losses. With secure property rights, the most attractive investment for some countries was simply leaving their oil in the ground. A former oil minister for Kuwait noted that his nation has the capacity to produce 5 million barrels daily but is lifting at a maximum rate of only 2 million barrels daily, endeavoring to spread the oil resources evenly over 80 years. This policy is rational for Kuwait given (1) secure property rights and (2) relatively unattractive investment opportunities. It is a rational policy independent of a cartel. Oil-producing countries, with the exception of Saudi Arabia, appear to take the price of oil in the long run as market determined and unaffected by their own behavior.

Table 4

OPEC COUNTRY OUTPUT AND MARKET SHARES, CLASSIFIED BY "CORE OF BALANCER COUNTRIES" AND "HIGH-POPULATION, HIGH-MOBILIZATION" COUNTRIES

	1974		1977	
	Output Per Day (000 bbls)	Market Share in OPEC (percent)	Output Per Day (000 bbls)	Market Share in OPEC (percent)
"Core of balancer" countries				
Saudi Arabia	8,481	27.6	9,200	29.5
United Arab Emirates	1,689	5.5	2,009	6.4
Kuwait	2,547	8.3	1,969	6.3
Libya	1,521	4.9	2,080	6.7
Iran	6,022	19.6	5,699	18.3
Iraq	1,975	6.4	2,265	7.3
Total	22,235	72.3	23,222	74.4
"Competitive fringe" countries				
Nigeria	2,256	7.3	2,097	6.7
Indonesia	1,375	4.5	1,685	5.4
Algeria	1,009	3.3	1,123	3.6
Ecuador	174	0.6	183	0.6
Gabon	202	0.7	222	0.7
Total	5,016	16.3	5,310	17.0
Unclassified countries				
Qatar	519	1.7	445	1.4
Venezuela	2,976	9.7	2,238	7.2
Total	3,495	11.4	2,683	8.6
Total OPEC	30,746			

Source: American Petroleum Institute, *Basic Petroleum Data Book.*

Given firmly established property rights and lower discount rates, one would expect reduced output growth rates and sharply higher prices. The record shows that from 1973 to 1977 Middle Eastern oil output increased at a compound annual rate of only 0.7 percent. This is in contrast to a 15 percent compound annual rate from 1970 through 1973 under insecure property rights and correspondingly high discount rates. As a consequence, crude oil prices in world markets (delivered in the United States) rose sharply, from approximately $3.25 per barrel in early 1973 to about $14.50 per barrel in 1977.[9]

Some real world complications must be introduced at this point. (1) Discount rates will differ from country to country reflecting each nation's opportunity costs of capital. (2) Future crude oil prices will be perceived differently. (3) The backstop technologies (competing energy sources such as shale oil, coal, nuclear fission and possibly fusion), which in the long run place upper limits on the price of crude oil, are likely to be evaluated differently by different analysts. (4) Member countries in OPEC have widely differing reserves extending for different periods into the future. Consequently, their evaluation of present values per unit will differ. (5) Finally, the optimizing principle itself in which independently competitive producers maximize the present value of their resource may be modified by political and strategic considerations. Choices between present and future production may be moderated as a result of pressures within the Arab community or by pressures from the United States. Variations in oil production are clearly part of the bargaining chips which countries may use to obtain concessions relative to military equipment or perhaps Middle Eastern politics. While the real world is pluralistic, whether model builders like it or not, there is still no need for a cartel theory to explain the 1978 price of crude oil. Given relatively secure property rights and low opportunity costs of capital for the "saver countries" in particular, the 1978 price of oil appears to be adequately explained by capital theory.

Price Effects of the Iranian Supply Reduction

The political turmoil in Iran reduced Iranian crude oil production from a peak of about 6 million barrels per day to a level too low to permit significant crude oil exports during the first quarter of 1979. The record of crude oil production by country is shown in table 5. This record shows that nearly all producing nations expanded oil production so that 58 percent of the Iranian supply reduction was offset.

---

[9]As 1977 opened, William Nordhaus in a *Wall Street Journal* article showed that this latter price was roughly consistent with rational expectations about the future value of oil.

Table 5

CRUDE OIL PRODUCTION BY COUNTRY

Country	1976	1977	1978 Jan	1978 Feb	1978 Mar	1978 Apr	1978 May	1978 June	1978 July	1978 Aug	1978 Sep	1978 Oct	1978 Nov	1978 Dec	1979 Jan
Saudi Arabia	8,579	9,206	7,758	8,355	7,705	8,010	7,258	7,495	7,421	7,154	8,382	9,308	10,252	10,406	9,800
Iran	5,883	5,663	5,292	5,530	5,602	5,611	5,704	5,763	5,804	5,805	6,053	5,542	3,494	2,371	445
Iraq	2,150	2,215	2,100	2,500	2,500	2,500	2,400	2,400	2,400	2,650	2,900	3,000	3,100	3,100	3,100
Venezuela	2,290	2,234	1,785	1,625	2,065	2,230	2,019	2,319	2,287	2,185	2,254	2,306	2,251	2,345	2,263
Libya	1,921	2,065	1,898	1,807	1,889	1,864	1,929	1,996	2,100	2,030	2,019	2,052	2,177	2,200	2,100
Nigeria	2,050	2,100	1,638	1,566	1,521	1,693	1,720	1,893	1,913	2,060	2,109	2,114	2,276	2,396	2,448
Kuwait	2,150	1,966	1,696	1,708	2,114	1,990	1,812	1,932	1,952	2,350	2,590	2,138	2,650	2,199	2,649
Indonesia	1,505	1,686	1,650	1,706	1,706	1,689	1,699	1,650	1,578	1,616	1,587	1,611	1,591	1,590	1,600
Total	26,528	27,135	23,817	24,797	25,102	25,587	24,541	25,448	25,455	25,850	27,894	28,071	27,991	26,607	24,405
Other OPEC	5,879	3,948	3,587	3,753	3,698	3,796	3,686	3,730	3,770	3,779	4,004	4,002	3,983	4,086	4,078
Total OPEC	30,407	31,083	27,404	28,550	28,800	29,383	28,227	29,178	29,225	29,629	31,898	32,073	31,774	30,693	28,483
United States	8,113	8,225	8,488	8,413	8,573	8,791	8,760	8,761	8,724	8,715	8,667	8,720	8,727	8,703	8,686
U.S.S.R.	10,364	10,920	11,115	11,185	11,209	11,315	11,327	11,360	11,444	11,495	11,534	11,650	11,700	11,665	11,650
U.K.	247	768	885	945	874	981	1,108	1,114	1,092	1,104	1,088	1,156	1,282	1,350	1,459
China	1,688	1,788	1,850	1,850	1,850	1,900	1,900	1,900	1,900	1,900	1,950	2,000	2,000	2,000	2,100
Canada	1,303	1,397	1,380	1,365	1,360	1,147	1,203	1,555	1,184	1,309	1,263	1,391	1,520	1,543	1,454
Mexico	801	981	1,107	1,120	1,100	1,139	1,151	1,167	1,205	1,239	1,267	1,297	1,325	1,366	1,390
Total	22,516	24,079	24,825	24,878	24,966	25,273	25,449	25,857	25,549	25,762	25,769	26,214	26,554	26,627	26,739
Other non-OPEC	4,342	4,607	4,812	4,830	4,894	4,910	4,904	4,855	4,944	4,996	4,859	4,840	4,893	4,861	4,972
Total non-OPEC	26,858	28,686	29,637	29,708	29,860	30,183	30,353	30,712	30,493	30,758	30,628	31,054	31,447	31,488	31,711
Total world	57,265	59,769	57,040	58,258	58,660	59,566	58,580	59,890	59,718	60,387	62,526	63,127	63,221	62,181	60,194

Source: *Oil and Gas Journal,* various issues.

From September 1978, when Iranian production was at its peak, to January 1979, Iranian production declined from 6.053 million b/d to a mere 0.445 million b/d, a drop of 5.608 million b/d. However, OPEC producers other than Iran expanded output by 2.193 million b/d and non-OPEC producers expanded output by 1.083 million b/d. Therefore the net reduction in global output was 2.332 million b/d or 3.7 percent.

Saudi Arabia, which produced 13.4 percent of world oil in September 1978, accounted for 43 percent of the gross increase in output. All of the major oil-producing nations shown in table 5 expanded production.

The foregoing comparison is based on September 1978 as the pre-Iranian crisis month. However, in that month OPEC production was at an uncommonly high level. If August 1978 is used as a base, then there is no significant change in world crude output relative to January 1979. When January 1979 is compared to one year earlier, we find that world production had actually increased by 5.5 percent. Or if average daily production for the first half of 1978 is used as the base, January 1979 production is higher by 2.6 percent.

The economic impact of the Iranian supply reduction should be viewed in both short- and long-run terms. In the short run, a minor net supply reduction is likely to produce a sharp price increase because the demand for crude oil is highly inelastic. A reduction in aggregate supply will lead to expected and actual price increases, in turn leading to some inventory buildup intended to capture inventory profits for intermediate processors. Consumers who expect rationing to be either instituted, or where it already exists to be further tightened, will attempt to increase their inventories, further stimulating price increases. This is often referred to as "driving on the top half of the tank."

But, from a capital theory viewpoint, there is no reason to expect a higher price trend over time for crude oil as a result of the Iranian supply reductions. A reduction in Iranian crude oil production in the present simply means that more Iranian oil will be available in the future than previously expected. Where Saudi Arabia and other oil-producing countries increase their oil production in order to offset some of the Iranian supply cutback, these countries must produce less in the future. Crude oil production is unlike lost wheat production or lost output through labor disputes. In those cases, once the losses have been sustained, the foregone output is permanently sacrificed. The Iranian crude oil remains in the ground for higher levels of future production; there is no change in long-run supply. To the contrary, if demand is reduced as a result of the short-term crisis, then the long-run supply-demand relationship is actually more favorable to lower future prices than previously expected.

When Iranian production is resumed, perhaps at a reduced rate, middlemen and consumers are likely to change their expectations. They would then anticipate an end to short-run price increases and expect price reductions from the "crisis" levels. This shift in expectations can in turn lead to a flow of oil out of inventories. Oil inventories are costly to hold. After short-run price adjustments complete their cycle, the trend in crude oil prices is likely to resume its pattern as determined by capital theory, possibly on a slightly lower trend line. Most producing countries will determine their production profile over time in such a way as to maximize their perceived interests, primarily the present value of their resources.

Gasoline shortages that appeared in March 1979 reflect the short-run supply forces discussed above. The short-run inventory demand factors are also part of the explanation for the observed shortages. However, United States goverment energy policy is similarly part of the problem.

Gasoline price controls were ineffective during the period 1977 into at least early 1978.[10] There were no shortages during this period. Government policy has attempted, through a system of mandatory price controls, to hold domestic crude oil and some oil product prices at artificially low levels. These controls have constrained refinery profit margins and have not allowed normal rates of return on capital for refinery investments, including investments in no-lead gasoline production and facilities to utilize high-sulfur crude oil. Thus, the price control system is part of the cause of United States gasoline supply problems which were developing in 1977-1978 and would have appeared, although with reduced severity, in the absence of any Iranian crisis.

Summary

Uncontrolled United States crude oil prices, after adjustment for inflation, have increased 140 percent from 1970 through June 1978. From 1975 onward, real prices of crude actually have declined 13 percent. These price movements have been explained in this paper as a rational application of modern capital theory combined with property rights theory. Cartel theory fails to explain output behavior of major crude oil-producing nations. Moreover, a cartel theory is unnecessary as an explanation of world crude oil price behavior in the 1970s.

The Iranian supply reduction in the winter of 1978-1979 produced sharp price increases and gasoline shortages in the United States. However, the Iranian problem is of short-run significance only. It should have no lasting effect on the long-run upward trend in oil prices. The price of crude oil, absent the Iranian problem, is seen to be a rational reflection of the

---

[10]Robert Deacon; "An Economic Analysis of Gasoline Price Controls" *Natural Resources Journal*, October 1978, 801-14.

price-limiting power of future oil production from huge oil shale reserves. The United States gasoline shortage appearing in March 1979 would have occurred as a result of government oil price control policies without the added impact of the Iranian problem.

6

OPEC: CARTEL OR CHIMAERA?

*René D. Zentner**

In the myth of Pegasus and Bellerophon, Proteus, the King of Argos, asked Bellerophon to slay the unconquerable Chimaera. The legend describes the Chimaera as a most singular portent – a lion in front, a serpent behind, and a goat in between – "A fearful creature, great and swift of foot and strong, whose breath was flame unquenchable."[1] In the myth, the Chimaera slew all who approached it on foot. But Bellerophon, mounted on the winged steed Pegasus, was able to soar above it beyond its fiery breath and dispatch the monster with arrows. What is important to us, however, is that, fearsome as the Chimaera was, it existed solely in Greek mythology.

Energy, too, has a mythology, which is only now beginning to be explored. One recent work has already addressed that mythology, dealing with such matters as the unfairness of oil prices, the relation of the Israeli question to world oil prices, and the effectiveness of the International Energy Agency (IEA).[2] In that work, Charles F. Doran explains how myths arise, pointing out that

---

* René D. Zentner is Manager of Corporate Studies of Shell Oil Company, Houston. His department conducts public opinion surveys and market research aimed at improving the quality of Shell's decision making. Prior to accepting his current assignment, the author held positions in Shell's patent and licensing organization, its transportation and supplies division, and its planning and economics group. He has a degree in engineering chemistry from Stanford University, as well as a J.D. degree from the University of California. Before joining Shell, Dr. Zentner was associated with E. I. duPont de Nemours & Company, Inc., with a San Francisco law firm, and with Tracerlab. He is a member of state and federal bars and was admitted to practice before the U.S. patent office. The coauthor of texts on radiochemistry and law, he has published numerous articles in legal and technical journals. Dr. Zentner has taught at the University of California and currently holds an appointment as Adjunct Associate Professor at Bates College of Law at the University of Houston.

Edith Hamilton, *Mythology: Timeless Tales of Gods and Heroes* (New York: Mentor Books, New American Library, Inc., 1940), p. 137.

[2] Charles F. Doran, *Myth, Oil and Politics: Introduction to the Political Economy of Petroleum* (New York: Free Press, 1977).

> Like O'Neill's protagonist, the Emperor Jones, who allowed mysterious little fears to become giant nocturnal figures, so a society embodies its fears, guilts and unfulfilled aspirations in myth, which in turn can imprison it. The variety of political ghosts and goblins that nations and other global actors fabricate is quite endless and not usually interesting unless a particular myth grows in size and plausibility to the extent that it begins to create mischief in the decision process.[3]

In the following discussion, I will explore one of the currently popular bodies of mythology existing in international affairs today: the myth of OPEC.

The realities of the Organization of the Petroleum Exporting Countries have been explored elsewhere, both in this volume and in the literature. There is, however, little comment in that literature on how OPEC is perceived. It is these perceptions – the dimensions of the OPEC myth – that this chapter will examine. In particular, I will deal with the following dimensions of the myth: (1) the positioning of OPEC in the national energy situation; (2) public perceptions of the relationship between energy prices and OPEC; and (3) the impact of OPEC on the United States. The data reported will be those drawn from such national surveys as those taken by Yankelovich, Skelly and White, Inc.; Louis Harris and Associates, Inc.; and Cambridge Reports, Inc.

Before turning to the discussion, however, it is worth asking just how extensive the American perception of OPEC is. In fact, public recognition of that Organization and its nature has arisen only comparatively recently, since the Arab oil embargo of 1973-1974.

OPEC had its twentieth anniversary in late summer of 1980. It was founded in September 1960 by five leading oil-exporting countries, three of which were Arab states – Iraq, Kuwait, and Saudi Arabia – and two of which were not – Iran and Venezuela.[4] Nevertheless, a search of the Gallup Polls beginning in 1960 revealed no reference to OPEC until after 1973. Similar results were obtained from a study of the indexes of the Harris Survey and a search of the files of Yankelovich, Skelly and White, Inc.

The reason for this scarcity is not hard to find. Until 1974, OPEC and its policies played a rather small role in international affairs. OPEC activities had been given little space in the press, especially in the American press. Space devoted to OPEC changed dramatically, however, beginning in 1974, with the emergence of OPEC's role in world oil pricing. This discontinuity in the amount of OPEC press coverage is illustrated by the data in table 1, derived from the Information Bank of the *New York Times*.

---

[3] Ibid., p. 5.

[4] Zuhayr Mikdashi, "The OPEC Process," *Daedalus*, fall 1975, p. 203.

Table 1

NUMBER OF CITATIONS IN THE NEW YORK *TIMES* INFORMATION BANK ON OPEC SINCE 1969

Year	New York *Times* (NYT) Only	All Citations	No. of Journals in Addition to NYT
1969	0	0	0
1970	4	4	0
1971	33	44	54
1972	68	104	65
1973	78	135	67
1974	265	457	67
1975	543	1,024	63
1976	273	584	75
1977	200	502	65
1978	153	287	66
1979	357	831	85
1980 (through April 28)	41	136	Not complete

Source: New York *Times* Information Bank.

While press attention to OPEC during its first decade appears modest, the 1973-1974 Arab oil embargo and the associated oil price increases by oil-producing nations brought the Organization into prominence: between 1973 and 1974, references to OPEC in the *New York Times* alone more than trebled. Since the major national surveys of public opinion tend to examine only relatively prominent issues on which the public has definite views, it is not surprising that little attention was paid to OPEC in such surveys until after the embargo.

It is by now generally established that the 1973-1974 change in world oil prices stemmed from oil-producing country decisions and that the oil-exporting nations, 13 years after OPEC was formed, had gained control of oil prices and production.[5] Nevertheless, the American public believed in 1974, and still believes, that the national energy problem arose from domestic rather than foreign origins. It is to those beliefs that I now turn.

[5] Robert Stobaugh and Daniel Yergin, eds., *Energy Future: Report of the Energy Project at the Harvard Business School* (New York: Random House, Inc., 1979), p. 28.

OPEC and the U.S. Energy Situation

The 1973-1974 oil embargo caused substantial shock in the United States. U.S. domestic oil production had peaked in 1970, and after that, the spare oil production capacity that had cushioned the impact of earlier supply interruptions was gone. By the fall of 1973, the U.S. energy economy had become heavily dependent upon imported oil. As a consequence, its domestic petroleum prices were increasingly influenced by the world oil price, which had been rising since the beginning of the decade. In 1973, oil imports amounted to 30 percent of U.S. oil consumption; by mid-1973, direct imports of Arab oil were exceeding 1 million barrels per day (b/d), up from less than half that amount 18 months earlier.[6] Thus the embargo struck two practically simultaneous blows: one political, revealing U.S. energy vulnerability, the other economic, in the sudden substantial increases in refined product prices.[7]

The American public had little understanding of the causes of the 1974 price changes and the associated brief period of product shortage. By an overwhelming majority, they blamed the oil companies and the U.S. government – the oil companies for the price increases and the government for allowing the price increases to happen. As recently as 1979, the oil companies remained the principal scapegoat, as table 2 indicates. The table

Table 2

INDUSTRIES/INSTITUTIONS THAT ARE TO BLAME FOR THE ENERGY PROBLEM

Thinking about the energy situation in general, which of these, if any, do you feel are mainly to blame for the energy problem? Pick as many as apply.

Industries/Institutions[a]	Percent Responding 1974	1976	1978	1979
Oil companies	72	63	64	73
Arab oil producers	40	49	57	67
The government	65	49	51	46
The public	24	33	39	39
Gas utilities	23	32	34	33
Electric utilities	21	33	36	32
Automobile manufacturers	26	29	30	31
Heavy industries	31	32	30	26

Source: Yankelovich, Skelly and White, Inc., *Corporate Priorities 1979* (New York, 1979), table D-2.6.

[a]Shows all mentions more than 20%.

[6] James W. McKie, "The United States," *Daedalus,* fall 1975, pp. 73-90.
[7] Ibid., p. 73.

shows the two principal characteristics of how the American public apportions responsibility for their energy concerns. First, of course, it illustrates the public's continuing distrust of oil companies. But it also reveals that public understanding of the nature of oil-producing countries is not clear. Throughout the survey data, "OPEC" is equated with "Arab," as we shall see.

Table 2 shows a steady rise in public recognition that foreign oil producers were also to blame for the national energy situation. This perception is confirmed by the data in table 3. Both tables reveal that oil companies and oil producers now share blame for the energy problem in the United States.

A similar conclusion was reached in mid-1979 by survey respondents who were asked about the reason for the occasional U.S. gasoline shortages occurring at that time. Table 4 indicates that half the population specifically blamed "big oil," saying that the supply problem was an industry conspiracy to drive prices up. Once again, OPEC and Arabs appear equated in the

Table 3

MOST IMPORTANT CAUSES OF THE NATION'S ENERGY PROBLEM, 1979

All of the following have been cited as possible causes of the energy problem. Which of these, if any, do you think are the most important causes of America's current energy problem? Pick as many as apply.

Most Important Causes (1979)	Percent
The actions of foreign oil-producing countries in raising prices	62
The actions of major oil companies, such as holding back supplies in order to boost prices	61
Waste of energy by consumers	58
Waste of energy by industries in their operations	52
The automobile industry is not doing enough to improve gasoline mileage standards	49
Natural resources are diminishing generally	43
The nation's failure to adequately fund research and development of alternative energy sources	42
The lack of a national energy policy	32
Delays in approving construction of energy facilities such as nuclear power plants	29
Overly strict environmental standards have restricted production of oil, gas, and coal	27
Price controls have reduced incentives to develop new domestic energy sources	22
None of these	1

Source: Yankelovich, Skelly and White, Inc., *Corporate Priorities 1979* (New York, 1979), table D-2.8.

Table 4

WHY GASOLINE SHORTAGES?

If there is a shortage of gasoline, why do you think it is happening?

Reason (1979)	Percent
Big companies or oil companies driving prices up, oil companies holding back for higher prices	50
Reduced supply, not enough to go around	7
People using too much, overconsumption, not conserving	8
OPEC, Arabs selling less oil	3
Not enough refineries or capacity to refine	2
Too much government restriction, regulation, or control, EPA, etc.	1
Government in general	1
Politics	1
President Carter	1
Situation in Iran	1
Foreign policy or foreign problems	1
Higher prices	1
Other	12
Don't know	9

Source: Cambridge Reports, Inc., *Cambridge Report 19* (Cambridge, Massachusetts, Second quarter, 1970), p. 210.

Table 5

THE MEANING OF OPEC

Can you tell me what OPEC stands for? (If "yes") What?

Meaning (1979)	Percent
Organization of Petroleum Exporting Countries or a close version	35
Oil producers, oil cartel	21
Arabs	4
Other / completely wrong answer	3
Don't know	37

Source: Cambridge Reports, Inc., *Cambridge Report 20*(Cambridge, Massachusetts, Third quarter, 1979), p. 251.

public view, but only by a minority of the public. In fact, the 1979 gasoline shortages appeared to result from the three-month cessation of Iranian oil exports, following the December 1978 Iranian revolution, rather than from either OPEC or Arab actions.

Nevertheless, since 1973 there has been a growing realization in the United States of the identity of OPEC and its role. In the late summer of 1979, about one-third of the American public could correctly explain the meaning of the acronym OPEC, and another quarter believed it had something to do with oil production or the Arabs. Table 5 reveals these perceptions but shows that 40 percent of the public still have no idea of what the acronym signifies. As one might expect, awareness of the meaning of OPEC increases steadily with income and education.

Most of the respondents who knew something about OPEC in the 1979 survey did identify, albeit vaguely, some aspect of OPEC's role in setting the price of oil, as table 6 shows. The themes of oil price controls are repeated by

Table 6

THE ROLE OF OPEC

(If respondent gave answer other than "don't know") Can you tell me what OPEC does?

Activity (1979)	Percent
Controls oil prices, sets prices on crude oil, they meet once a year and set gas and oil prices	30
They are a cartel, a cartel that sets oil prices, a group of countries that combine their oil and sell at the same price	10
Controls the output of oil, controls sales of oil	6
Regulates shipment and production of oil	6
Sells oil to the U.S.	5
Sells oil to the world, sells oil internationally	5
Produces oil, oil-producing countries	5
Controls oil prices for the world, sets prices abroad	4
Keeps raising prices, puts prices higher and higher	4
Screws America, puts gun in our ribs, rips off America	3
Exports oil to foreign countries	3
Controls oil prices for the U.S., sets U.S. oil prices	2
Controls output of oil to U.S., controls amount of oil we get	1
Charges excessive prices	1
Imports oil	1
Acts as a front for American oil companies to raise prices, in collusion with American oil companies	1
Controls the world economy with their oil	1
Other	1
Don't know	11

Source: Cambridge Reports, Inc., *Cambridge Report 20* (Cambridge, Massachusetts, Third quarter, 1979), p. 253.

40 percent of the respondents who knew something of OPEC and recognized the role OPEC plays in setting world oil prices.

That OPEC's function is not beneficial to the United States is perceived by almost two-thirds of the public. During mid-1979, Cambridge Reports asked Americans whether they thought the Organization's activities were good or bad for this country. By a margin of more than four to one, respondents felt that the United States was harmed by its purchases of oil from OPEC (see table 7). It is interesting, however, to observe that almost a quarter of all respondents expressed no view on the issue, showing that a large proportion of the population was either ignorant of OPEC influence or indifferent to it.

The foregoing data show that the nature and role of OPEC in the world energy situation is not very well understood in the United States. Nevertheless, Americans aware of OPEC recognize that it has some influence on world oil prices, and that such influence is not beneficial to this country. It is now appropriate to further examine American perceptions of the impact of OPEC on energy prices.

Relationship Between OPEC and Energy Prices

In examining public perceptions of energy prices, it is important to understand how the national energy situation is viewed. Surveys conducted by Yankelovich, Skelly and White, Inc., from 1976 to 1979 reveal that the energy crisis is principally seen as one of price (table 8). Over the four-year period reported, perception of the energy problem as one of predominantly a price problem rose from 63 percent of the public to 71 percent, while energy prices themselves also rose.

As the previous tables have shown, there is some awareness both that the United States must import some of its oil and that the price of imported oil is

Table 7

OPEC — GOOD OR BAD FOR THE U.S.?

OPEC stands for the Organization of the Petroleum Exporting Countries and consists of the major nations in the world that sell oil to the U.S. and other countries. In general, from what you've heard, do you think the existence of OPEC has been good for the United States or bad for the United States?

	Percent Responding in 1979
Good	14
Don't know	23
Bad	63

Source: Cambridge Reports, Inc., *Cambridge Report 20* (Cambridge, Massachusetts, Third quarter, 1979), p. 253.

Table 8

CONCERNS OVER ENERGY PROBLEM

People talk a lot about the energy crisis. Which, if any, of these things are of particular concern to you with respect to the problem of energy?

Concerns	Percent Responding 1976	1977	1978	1979
High price of energy	63	65	70	71
Dependence on foreign countries..................	53	58	63	62
Power of the oil companies ...	53	48	56	51
Shortages occurring in next two years	39	46	40	48
Shortages occurring before the end of century	37	38	40	36

Source: Yankelovich, Skelly and White, Inc., *Corporate Priorities 1979* (New York, 1979), table D-2.9.

in some fashion controlled by OPEC. In this regard, public awareness of the proportion of U.S. oil supply that is imported is not great. During 1979, the United States imported 45 percent of its oil, but only 30 percent came directly from OPEC countries.[8] As table 9 reveals, however, only about one-half of

Table 9

PERCENTAGE OF U.S. OIL IMPORTS

What percentage of oil that America used last year was imported from foreign countries?

Import Proportion	Percent Responding 3rd Quarter 1976	3rd Quarter 1977	2nd Quarter 1979
Less than 10%	4	2	4
10-20%	10	6	8
21-33%	14	14	11
34-50%	24	24	23
51-75%	16	25	22
76% and above	4	5	5
Don't know	28	23	28
% Imported previous year[a] ...	37	42	48

Source: Cambridge Reports, Inc., *Cambridge Report 19* (Cambridge, Massachusetts, Second quarter, 1979), p. 203.

[a]Actual percent imported; U.S. Department of Energy, Energy Information Administration, *Monthly Energy Review*, March 1980, p. 16.

[8] United States, Department of Energy, Energy Information Administration, *Monthly Energy Review*, March 1980, p. 16.

the American public understands that this country imports a substantial proportion of its oil supply; about one-fourth could not even estimate the extent of national oil imports. Nevertheless, there is considerable public recognition, as table 3 indicated, that U.S. petroleum prices are affected by OPEC.

That oil price increases have been unjustified was the view of about two-thirds of the Americans surveyed in 1977. Table 10 shows that nearly five times as many Americans believed the OPEC price increases were a "rip-off" as believed them to be justified. As in previous tables, however, table 10 also shows that about one-fifth of the public did not have an answer to the question.

Despite these attitudes, Americans are not quick to link domestic petroleum product prices with foreign influences. As table 2 showed, the public regards the energy problem as principally caused by the oil industry. While domestic refined product prices are rising as a result of rising world oil prices, less than one-fifth of the public was able to make that connection, as table 11 indicates. In late 1979, Cambridge Reports asked purchasers of home heating oil why the furnace oil price had risen. Over half of the respondents attributed the price increase to either oil company greed or to oil shortages.

The reasons behind OPEC increases in the world oil price are not well understood by the American public. Very few Americans believe that world oil prices have been increased because of concerns over resource exhaustion or for economic development. Most frequently mentioned as the reason behind OPEC price increases is sheer greed or simply the desire for more revenue. These conclusions stem from the responses such as those to a question asked the American public in 1977 by the Opinion Research Corporation (see table 12).

Table 10

OPEC PRICE INCREASES JUSTIFIED

Since 1973, the nations that are members of the OPEC cartel — the Organization of the Petroleum Exporting Countries — have raised the price of crude oil about 400 percent. Some people say this represents a justifiable move by a group of largely underdeveloped countries to get a fair price for their natural resources from the largely industrial nations that buy their crude oil. Others say the price increases have been a "rip-off" that take billions of dollars out of the pockets of consumers all over the world, but particularly in the United States. Which is closer to your view?

Third Quarter, 1977	Percent
Increases are justified	14
Don't know	19
Increases are a "rip-off"	67

Source: Cambridge Reports, Inc., *Cambridge Report 12* (Cambridge, Massachusetts, Third quarter, 1977), p. 171.

Table 11

REASON FOR PRODUCT PRICE INCREASE

As you may know, the average price of home heating oil is currently about 40¢ more per gallon than it was last year at this time. What do you think is the main cause of this price increase?

Main Cause (4th Quarter, 1979)	Percent
Oil companies' greed / increasing their profits / ripping us off	33
Oil / fuel shortage, not enough oil to go around	23
OPEC countries raising prices	9
Oil imports, foreign oil	8
High cost of oil	4
Inflation, high wages	4
Decontrol of oil prices	2
Problems in Iran or Middle East	2
It's all going to cars, trucks, or airplanes	1
Other	3
Don't know	10

Source: Cambridge Reports, Inc., *Cambridge Report 21* (Cambridge, Massachusetts, Fourth quarter, 1979), p. 203.

Table 12

REASON FOR OPEC OIL PRICE INCREASE

At their recent meeting, the petroleum-exporting countries did decide to raise the price of oil, although there was disagreement as to how much the price increase should be. Why do you think the petroleum-exporting countries are raising their prices?

Reason (1977)	Percent
To make more money, increase profits, greed	36
In response to increased demand for oil	25
Need for more money, to meet inflation, increasing costs	15
To increase or demonstrate their strength and power in the world	12
To get more money for their domestic programs	11
Need to conserve energy, petroleum is getting scarce	3
Need more money to buy military arms	2
Other nations are dependent on them for oil	2
To increase political pressure against Israel	2
They can get away with it	1
Other answers	7
Don't know	18

Source: Opinion Research Corporation (Princeton, New Jersey), *Public Opinion Index*, vol. 25, no. 2 (January 1977), p. 2.

Once again, it is interesting to observe that in 1977 almost two out of 10 Americans had no idea what motive underlay OPEC oil price increases. This comparatively large proportion reinforces points made above about the substantial lack of public perception and understanding of OPEC as a force in petroleum price formation.

The data in table 12 are from a survey conducted in January 1977. That these views persisted in subsequent years is evidenced by data collected in 1978 and 1979 by Louis Harris and Associates, Inc. (shown in table 13). What is interesting about these data is the 10 percentage-point decline in the "undecided" category, accompanied by an almost corresponding increase in the "agree" column. This change suggests an increasing awareness in the United States of the producer countries' role in oil price increases.

Thus, even though the national energy problem is seen as one of price, and despite public indignation over OPEC price increases, there is not widespread recognition that U.S. petroleum product prices are influenced by OPEC price behavior. Moreover, there is little public understanding of producer countries' reasons for raising prices. It is in the light of these perceptions, or lack of them, that I now turn to the question of how the public feels about U.S. policies toward OPEC.

OPEC and U.S. Policy

It is now appropriate to examine how public perceptions of OPEC affect public beliefs about national policy. Before doing so, however, a look at why Americans are concerned about OPEC is informative. As we have seen, although there is not widespread understanding of the OPEC role in product prices, the public believes that producer price increases arise from selfish motives. During 1979, more than two-thirds of Americans also believed that OPEC oil price increases would damage the U.S. economy (table 14). About

Table 13

PRODUCING-NATION INTEREST IN NATIONAL INCOME

The Arab oil-producing countries are more interested in their own national income and development than in bringing in more oil at reasonable prices to meet the world's energy needs.

	Percent Responding	
	March 1978	April 1979
Agree	77	89
Disagree	7	4
Not sure	16	6

Source: Louis Harris and Associates, Inc. (New York), *Harris Perspective – 1979,* vol. II, report no. 56, p. 64.

one-fourth of the public thought the price increases would seriously damage the national economy and another 43 percent thought damage would be perceptible. Only one-fifth of the public thought damage would be negligible or non-existent.

Coupled with this concern over the effects of OPEC price increases is a feeling that U.S. oil companies are acting in the producing countries' interests rather than in the national interest. In 1979, more than half the public perceived such a conflict of interest, as table 15 reveals. Three times as many Americans saw oil companies acting in OPEC's interest than saw them acting in the best interests of the United States. There are thus two agencies in-

Table 14

EFFECT OF OPEC PRICE INCREASES ON U.S. ECONOMY

Recently, the Organization of the Petroleum Exporting Countries — OPEC — announced a major increase in the price of oil. Do you think that increase will seriously damage the United States economy, somewhat damage it, only slightly damage it, or will it not really affect the economy at all?

Second Quarter, 1979	Percent
Seriously damage economy	26
Somewhat damage economy	43
Only slightly damage economy	16
Not really affect economy at all	4
Don't know	10

Source: Cambridge Reports, Inc., *Cambridge Report 19* (Cambridge, Massachusetts, Second quarter, 1979), p. 204.

Table 15

U.S. OIL COMPANIES AND THE NATIONAL INTEREST

Obviously, American oil companies buy much of their oil from the OPEC countries. When they do this, do you think that they generally act in the best interests of the United States or do they act more for the OPEC countries?

Third Quarter, 1979	Percent
Best interests of the United States	18
Don't know	27
Act more for OPEC countries	55

Source: Cambridge Reports, Inc., *Cambridge Report 20* (Cambridge, Massachusetts, Third quarter, 1979), p. 254.

volved in the assault by OPEC policies on the American economy: OPEC itself and the American oil companies who are seen as acting on OPEC's behalf.

There are consequently two kinds of policy responses: one against the oil-producing nations and the other against the oil companies. For some years, there has been extensive public support for sanctions aimed at producer countries. More than two-thirds of the public has supported partial or complete cessation of trade with OPEC members. Less than a quarter of the public would continue trade without sanctions (see table 16).

In the same survey, the question of what particular goods shall be sanctioned was also explored. Food and military arms were the goods most frequently mentioned. These data indicate that Americans feel our displeasure with OPEC price increases can be shown through economic reprisals (table 17).

There is much less consensus on how the United States should deal with American oil companies who purchase oil from OPEC members. To date, there has been no serious suggestion that the United States discontinue purchases of OPEC crude oil. One proposal for making OPEC more responsive to U.S. interests has been to make the federal government the principal purchaser of foreign crude oil and thus replace the U.S. oil companies who now deal with producing nations. Observers of the effectiveness of U.S. government intervention in energy markets have expressed doubts that such an approach would be effective. Proponents argue that the method would assure that the interests of the American people would be put before oil company profits. Only a plurality of the public endorsed the proposal in 1979; almost as many Americans disagreed with the proposal, and more than one-fourth expressed no view on the question, suggesting that better solutions await discussion (see table 18).

Finally, there has been widespread dissatisfaction with President Carter's responses to OPEC. In an April 1979 study of public attitudes toward actions of the Carter Administration in international affairs, seven out of 10 Americans said they believed the administration was "too soft" in its dealings with OPEC (table 19). As world oil prices continue to rise, this position is likely to harden.

The Cartel or the Chimaera?

From an examination of the foregoing data, the first thing that can be said about the American public's perceptions of OPEC is that those perceptions do not yield a very clear view of that body. To the extent it is seen at all, OPEC is seen as having a subordinate role in energy and energy price issues, and having only a secondary influence on U.S. energy or economic affairs. The principal player in affecting those issues and influencing those affairs, at

Table 16

U.S. RESPONSE TO FUTURE OPEC OIL INCREASES

If the petroleum-exporting countries continue to raise their prices, should the United States respond by continuing to trade as we do now, by refusing to sell only certain goods to these countries, or should we refuse to sell all American goods to these countries?

U.S. Responses (1977)	Percent
Refuse to sell *certain* goods	43
Refuse to sell *all* goods	25
Continue trade	22
No opinion	10

Source: Opinion Research Corporation (Princeton, New Jersey), *Public Opinion Index*, vol. 25, no. 2 (January 1977), p. 3.

Table 17

SPECIFIC GOODS WHOSE EXPORT TO OPEC NATIONS SHOULD BE RESTRICTED

Specifically, what goods should we refuse to sell the oil-exporting countries if they raise oil prices?

Goods (1977)	Percent
Food	30
Military weapons and supplies	27
Things they really need	13
Industrial machinery and supplies	10
Nuclear plants, equipment and supplies	6
Transportation equipment and supplies	6
Farm equipment and supplies	4
Clothing	2
Luxury items	2
Medicine; hospital equipment and supplies	1
Other	6
No opinion	27

Source: Opinion Research Corporation (Princeton, New Jersey), *Public Opinion Index*, vol. 25, no. 2 (January 1977), p. 4.

Table 18

FEDERAL GOVERNMENT AS PURCHASER OF OPEC OIL

Some people have argued that the U.S. should forbid the oil companies from dealing with OPEC directly and should have the government buy all the oil we import. They say the government would be more inclined to try to get a good price for the country than the oil companies are. Do you generally agree or disagree with this view?

Third Quarter, 1979	Percent
Agree	41
Don't know	28
Disagree	31

Source: Cambridge Reports, Inc., *Cambridge Report 20* (Cambridge, Massachusetts, Third quarter, 1979), p. 254.

Table 19

ADMINISTRATION EFFECTIVENESS IN INTERNATIONAL AFFAIRS

Generally, do you feel that the Carter Administration has been too hard, too soft, or just about right in their dealings with:

	Percent Responding			
	Too hard	Too soft	Just right	Don't know
OPEC	3	70	21	6
Arafat and PLO (Palestine Liberation Organization)	2	57	27	14
Palestinians	5	47	37	11
Russians	2	44	49	6
Chinese	2	35	57	6

Source: Yankelovich, Skelly and White, Inc., *Corporate Priorities 1979* (New York, 1979).

least in the perceptions of most Americans, remains the American oil industry.

To the extent that the American public sees OPEC at all, it sees it murkily. Thus, "OPEC" and "Arab" appear to be employed interchangeably in the common language of the energy debate. The equating of the two terms is not especially surprising since Arab members have been among the most vocal and visible of the producing nations. Moreover, oil from Arab exporting nations forms the largest proportion of U.S. oil imports from OPEC or non-

Table 20

PROPORTION OF ARAB OIL IN U.S. DIRECT OIL IMPORTS

Source	Total 1979 Imports	
	Million Barrels/Day	% of Total
Arab OPEC members	3.02	36
Total OPEC	5.53	67
Total non-OPEC	2.75	33
Total imports	8.28	100

Source: U.S. Department of Energy, Energy Information Administration, *Monthly Energy Review*, March 1980, pp. 30-33.

OPEC sources, as table 20 reveals. The 3 million b/d of oil from Arab OPEC members amounts to 55 percent of total OPEC imports and exceeds the 2.75 million b/d from non-OPEC sources.

As to whether the public perceives the Organization as a cartel is less clear. Certainly "cartel" is one of the terms frequently applied to OPEC by government spokesmen, the press, and, as tables 5 and 6 *supra* indicate, by a comparatively small proportion of the public. Whether or not OPEC has been a cartel, or will be a cartel, is beyond the scope of this chapter. The question has been addressed in part by Professor Doran in his recent treatment of myths associated with the political economy of petroleum; Doran began, however, by treating OPEC as though it were a cartel, and then examined its potential cohesiveness.[9] The foregoing data suggest, however, that in the public mind, "OPEC" and "cartel" are equated only to a modest extent.

How, then, does the American public see OPEC? It sees it principally as a seller of oil to the United States and to the world, but also as an organization which, in selling that oil, sets and raises prices for unworthy reasons. Moreover, it believes that OPEC pricing policies are damaging to the U.S. economy and require retributive measures. From its incomplete perception of OPEC, the American public has, like the ancient Greeks, fashioned a chimaera. The new chimaera has an Arab head, a rapacious heart, economic fangs, and cat's paws in the shape of American oil companies.

Considerable national interest and attention is currently being devoted to ways in which the United States can slay this chimaera; since 1974, this country and other members of the International Energy Agency have been searching for a statesman Bellerophon and a policy Pegasus. So far, we have found

9 Charles F. Doran, op. cit., chapter 6. Compare with Salah El Serafy, "The Oil Price Revolution of 1973-1974," chapter 7 in this book, p. 125.

neither the monster killer nor his winged steed, though there have been many candidates. In the excitement of the hunt, however, the central question has not been extensively discussed. We began this discussion by pointing out that, despite its many fearsome attributes, the chimaera was only a myth. The central question, then, is whether we are seeking to kill a monster that does not exist.

7
THE OIL PRICE REVOLUTION OF 1973-1974

*Salah El Serafy**

Introduction

It is now five years since the oil price revolution took place in 1973-1974, and we are all the wiser about its implications. A few issues have clarified, but many others remain obscure. However, the present seems to be a good time to take stock of the issues we understand better, and see if we can identify new directions for a better future, not just for the oil producers but for everybody in our small interdependent planet. This paper will be structured around three main interrelated topics, namely, price, the balance of international payments, and finally, development policy (and its obverse, the levels of oil production) in the oil-producing countries.

Price

Since 1973 our perception has improved of the factors determining the price of oil, not so much because of any progress in economic analysis, but simply on account of the gaining in currency of the notion that petroleum is

*The author, who holds degrees in economics from Oxford (D. Phil.) and from the Universities of London and Alexandria (Egypt), has been a staff member of the World Bank since 1972, successively in the South Asia Region, the Europe, Middle East and North Africa Region, and the Operations Evaluation Department. Previously Dr. El Serafy served as Project Director, The Economist Intelligence Unit, London; Associate Professor of Economics, Alexandria University; Research Fellow, Center for Middle Eastern Studies, Harvard University; Research Fellow, Department of Economics, Harvard University; and Advisor to the Government of Libya under the auspices of the United Nations Development Program. He has conducted research, led teams and published works in the areas of resource development, project analysis, economic and social development, commodity markets, national accounts, and macro-economic forecasts. This article reflects solely the views of author and not necessarily those of the World Bank.

an exhaustible asset, not just an ordinary good. In competition its price should not be equal to the marginal cost of its extraction (often erroneously referred to as production). But, being of limited supply, its price should contain a scarcity rent, deriving from the price of its substitutes. Since supply is finite, it follows that extraction results in increased scarcity; the scarcity rent therefore should rise over time.

Analysis of equilibrium in the asset market indicates that the price of oil should rise over time to allow the scarcity rent contained therein to increase exponentially at the current rate of interest. The price will attain equilibrium only when the "producers" become indifferent between leaving a barrel of oil in the ground to appreciate in the future as its scarcity increases or extracting it and investing its value at the current interest rates. Over time the increase in the price of oil will discourage demand and keep it equal to the dwindling supply until eventually demand will have shrunk to the ultimate single barrel that remains and the world's stock of oil is exhausted. Should the prevailing interest rates, however, happen to be higher than the social rate of time preference, it would be to the advantage of the oil producers to accelerate its extraction and oil would be depleted faster than society truly wants; alternatively it would be conserved longer (i.e., depleted more slowly) than the true preferences of society would warrant.[1]

One advantage of a configuration like the above is that it brings into play the longer-term forces of supply and demand which are the relevant forces for this kind of analysis. And in thinking of the prices of alternatives, which determine the level of scarcity rent, the proper prices are the long-term supply prices, and these could in turn be dependent on the price of petroleum, including gas, which is the source of much of primary energy produced at present. Although the price of oil, even today, remains below the price of alternative energy sources, changes in technology, partly the product of the prevailing relatively high energy prices, could bring about a reduction in energy prices over the medium or longer term. Higher petroleum prices now could lead to lower supply prices for substitutes (increased supply) in the longer run; low petroleum prices would delay the development of substitutes and therefore contribute to higher substitute prices in the future. There are of course optimists and pessimists about

---

[1]Cf. Robert M. Solow, "The Economics of Resources or the Resources of Economics," *American Economic Review*, May 1974, who recalled the analysis of H. Hotelling of "The Economics of Exhaustible Resources," *Journal of Political Economy*, April 1931. In a competitve market where the demand curve facing individual producers is horizontal, the scarcity rent would be the difference between marginal cost and price. Where the demand curve is downward sloping, the scarcity rent would equal the rate of marginal profit, or the difference between marginal cost and marginal revenue.

likely progress in technology in this field. But, after the initial "optimism" that foresaw a collapse of the 1973-1974 oil prices, realism has tended to set the long-term price of substitutes quite high — much higher than was thought possible before.

Of course if we knew the future of technological change we would be able to set the long-run price of substitutes correctly, and from there would set the price of petroleum correctly and see to it that it should rise year after year to allow the scarcity rent to grow exponentially at the current rate of interest. Instead, we have the present state of affairs where the producers claim that current oil prices are too low, and the consumers, or at least some of them, insist they are too high.

Although we cannot know for certain which is right, it is becoming clear that current prices are probably not high enough to bring forth significant supplies of substitutes for the next 10 or perhaps 15 years. Be that as it may and with the benefit of hindsight, there is little doubt that the current levels of petroleum prices are more in harmony with the state of its scarcity and the supply prices of its substitutes that the exceptionally depressed prices of less than $2 a barrel for crude oil than prevailed for over two decades, with a declining trend until 1972.[2]

This, however, is past history; yet one could add that those depreseed prices in the past were not imposed on the industry out of wickedness, but resulted from the interplay of forces that mistook the cost of extraction for the cost of production and, perhaps understandably, ignored the interest of the host countries in having a say in the rate at which their exhaustible resources were being depleted. It must be conceded that many of these countries were themselves not aware of the wealth under their own soil until the oil companies had explored for and located it and had sunk huge investments for its development. And it was indeed a fortunate developing country that attracted the attention of the oil companies and secured royalties to help finance its development. The price of oil may have been depressed by the standards of today or even by other objective standards, but nobody dreamed in the sixties that it should or could reach the heights it did in 1974. The fact is that the then prevailing and indeed falling low prices induced the industrialized countries to build a way of life around cheap petroleum which by the early seventies had become the source of three-quarters of OECD energy. The modes of travel and living, the commuting to work and the rise of the automobile, the rapid decline of coal (as a fuel and as an industry able to attract and retain workers), and the slow search for clean and replaceable sources of energy were all the products of low petroleum prices So when the price revolution of 1973-1974 occurred,

[2]See Appendix, table 1, for a series of oil prices in current and constant values.

the forces to resist it were too weak to unravel it. In a sense, therefore, the low energy prices in the fifties and sixties were the prelude and the necessary condition for the higher prices that have prevailed since 1974.

That is not to say that the situation prevailing before the oil price revolution of 1973-1974 had been stable; sooner or later it would have occurred to the principal consumers of energy themselves that such a state of affairs could not last. Even if the oil producers' preferences were to continue to be disregarded, the consumers themselves would have eventually had to regulate extraction in accordance with their relative preferences as between immediate and future needs, and prices would have had to rise. It was only on account of the abruptness and magnitude of the price change in 1973-1974 that so many passions were raised.

The price levels established in 1973-1974 have largely stuck. They have failed to bring forth the deluge of energy oversupply which many people predicted at the time. Nor have they induced significant shifts of demand away from this very handy source of energy. Despite some fluctuations in the market, the 1973-74 price levels have stood the test of time, and, in view of the projected growth of demand, practically all serious analyses now indicate a probable rise in the real price of oil, particularly after 1985. Between now and 1985, especially under the influence of some significant additions to oil supplies in the North Sea, Alaska, Mexico, and a few less-developed countries, the real price for oil may not be subjected to intense upward pressure. But the pressure would greatly increase in the late eighties. Even by 1985 OPEC would be "required" to increase its production by between 4 and 12 million barrels per day for the 1977 real price of oil to remain constant.[3] Doubt about the oil producers' willingness to expand their future production to fill in the projected gap has prompted Walter Levy recently to ask:

> May we not be headed, within the next decade at most, for far more serious upheavals, in which the OPEC producing countries can — and may, if for no other reason than to protect their vital self-interest — completely reassess their present internal development policies, which, in turn, may drastically affect their policies on the supply and price of oil?[4]

Recent events in Mexico have also reinforced this notion: that unless the terms offered to the oil producers are attractive enough, they will be forced to regulate their production according to their own development requirements, which may not coincide with the energy needs of the

---

[3]Walter J. Levy, "The Years That the Locust Hath Eaten: Oil Policy and OPEC Development Prospects," *Foreign Affairs*, winter 1978-1979.

[4]Ibid.

consumers.[5] This point will be given greater attention in a subsequent section of this paper.

International Payments

In 1972 the value of world trade in oil was about $25 billion, equivalent to some 6 percent of world merchandise trade and about half of 1 percent of world income. By 1976 the value of the oil trade had risen to about $130 billion, accounted for some 13 percent of world merchandise trade and some 2 percent of world income.[6] The commotion caused by the oil price revolution of 1973-1974 contrasted with the modest place it occupied in world trade and relative to world income. The violence of the reactions to the price change stemmed, however, from the insecurity of supplies generated by the Arab embargo, the mistaken belief that the new prices, unlike the older ones, had no economic justification behind them, and lastly from the imbalance created in international payments by the new prices. Inordinate attention was to be given to this last point, almost to the neglect of all else.

In 1972 many of the densely populated oil producers were in deficit on current account, resorting to international borrowing to supplement their domestic resources in pursuit of economic and social development. Some of them, such as Nigeria and Indonesia, were also recipients of concessional aid from the richer countries. The aggregate international reserves of all the oil producers were modest and were surpassed by a number of individual country reserves including Germany and Japan. The price revolution of 1973-1974 upset this pattern and converted the surpluses of the net oil importers into deficits. The world held its breath in anticipation of a new situation in which the industrialized countries would be transferring a significant portion of their hard-earned income to a few primary producers. These producers would accumulate rising reserves, since their ability to import, alias absorptive capacity, was clearly limited. The rise of reserves would in turn generate income for the oil producers to add to the balance-of-payments burden of the oil consumers whether rich or poor. It was indeed a scenario which raised emotions high to the point of military threats being voiced against OPEC members.

In retrospect neither the surpluses nor the ever-rising reserves

---

[5] The clear implication of this is that the current prices for oil are too low as compared with the long-term equilibrium prices referred to earlier in this paper. See also the Appendix entitled "On OPEC As a Cartel."

[6] Based on trade data in International Monetary Fund (IMF), *International Financial Statistics* (Washington, D.C.: IMF) and on data in the World Bank, *Atlas* (Washington, D.C.: The World Bank).

materialized in the way the forcasters (and they were many) had predicted. The growth of imports of OPEC members since 1973 has been astonishing. Merchandise imports of the major oil-exporting countries increased in nominal terms by 44 percent in 1973, 69 percent in 1974, 60 percent in 1975, 22 percent in 1976, and 36 percent in 1977.[7] By contrast, their exports reached a plateau in 1974 and nominal growth since then has been in the neighborhood of 6 percent per annum. Their net service imports soared as payments for tourism, freight, workers' remittances and others exceeded all expectations. A major program of aid was also mounted, and large capital transfers for assistance and investment were made both in developing as well as developed countries. The oil nations made considerable contributions to international and regional development funds and cooperated in cofinancing with other donors a larger number of projects throughout the world. Their current accounts began to show diminishing surpluses: $68 billion in 1974; an average of $35 billion a year in the period 1975-77; an estimated $9 billion in 1978. This last-mentioned figure compares with an estimated 1978 current account surplus of $8 billion for Italy, $13 billion for Germany, and $17 billion for Japan. Thus the aggregate international reserves of this oil group decelerated in growth, and by end-November 1978 reserves amounted to 46.4 billion SDR — only 13 percent higher than the reserves of Germany.[8]

As to the nonoil developing countries, these began as a group to encounter large current account deficits after 1972. Their aggregate deficits reached a maximum in 1975, but have tended to diminish in later years, with the possible exception of 1978. The oil price increase has certainly been behind the deterioration in their external payments. However, since the energy needs of the low-income developing countries are on the whole modest, the impact of the oil price increase on their balances of payments therefore has been limited. More important, perhaps, was the effect of the general deterioration of their terms of trade, caused in particular by the leap in the price of their imports of manufactured goods and by the depressed world demand for their primary exports. While the various programs of aid and special arrangements, such as the IMF Oil Facility, did in general alleviate many difficulties, there have been a few cases of hardship, especially in nations where the aid received was nonexistent or otherwise too little to offset the damage done to their economies. In sum, the impact of the 1973-1974 oil price levels on

---

[7]That is, Algeria, Indonesia, Iran, Iraq, Kuwait, Libya, Nigeria, Qatar, Saudi Arabia, United Arab Emirates, Venezuela, and the non-OPEC state of Oman. See IMF, *International Financial Statistics*.

[8]Ibid., and table 2 in the Appendix.

international payments and international liquidity turned out to be different from the disaster that had been foreseen. Here it is interesting to note that Japan, perhaps the most dependent industrial power on imported fuel, got over its balance-of-payments difficulties quite early. On current account (including official transfers), Japan's 1974 deficit was almost eliminated in 1975, and a surplus of $3.6 billion appeared in 1976. In that year Italy and the United Kingdom also began to have surpluses on current account.

Some World Repercussions

In retrospect the world seems to have adjusted fairly well to the upheaval in the oil market in 1973-1974. This is not to say that 1974-1975 was not a turning point for the world economy, for in many respects it was. In the decade prior to 1972 the industrialized economies had grown in real terms by about 4 percent per annum (Japan leading the way by over 10 percent) and the less-developed economies by about 6 percent. Growth of both groups accelerated in 1973 to about 6 and 7 percent, respectively, but has declined since. The world experienced a severe recession in 1974-1975 with inflation combined with stagnation. Improvement in activity since then has been slow, with the industrialized economies averaging 3-4 percent per annum real growth, and the less-developed countries 4-5 percent. Underutilization of resources, particularly in manufacturing industry, remains severe in the industrialized countries, whereas it had been quite modest prior to 1972. Inflation has become a structural problem. In the decade prior to 1973 the GDP deflator in the industrialized countries increased on average by 4 percent annually, and consumer prices in the less-developed countries (excluding the more volatile ones) by 6-10 percent. After two years of pronounced inflation in 1974-1975, inflation seems to have settled to a level about 7 percent per annum in the industrialized countries and at 10-25 percent (again excluding extreme cases) in the less-developed world. The "world economic order" has also been plagued since the early seventies by continuous and often violent changes in exchange rates.[9]

As to the terms of trade, the oil producers made considerable gains in 1973, and massive gains still in 1974; afterwards and taking changes in

---

[9]The IMF is the source of most of the measurements cited in this article on the terms of trade, balance of payments, inflation, and growth. See various issues of the *IMF Survey*; the Communiques of the Intergovernmental Group of Twenty-Four on International Monetary Affairs (Ministerial Meetings); the IMF *Annual Reports*; Joint Press Conference by the Chairman of the Interim Committee Twelfth Meeting and the IMF Managing Director, March 7, 1979; the *Balance-of-Payments Yearbook* and *International Financial Statistics*.

exchange rates into consideration, they have lost ground in every year, with the greatest losses experienced in 1975 and 1978. Because oil is relatively less important in their economies, the industrialized countries' deterioration of terms of trade in 1973 and 1974 was on a much reduced scale, and so were their gains in 1975 and 1978. The less-developed nations improved their terms of trade in 1973, lost ground in 1974 and 1975, and appear to have gained a little in 1976 and 1977, to lose again in 1978.

Whether or not it was the oil price revolution of 1973-1974 and its repercussions that triggered the recession, intensified the inflationary tendencies and caused or accentuated exchange instability is not important for the purpose of this article. Although the roots of all these ills extend back in time to well before 1973, there is little doubt that the drastic and abrupt change in the oil terms of trade in 1973-1974 bared economic weaknesses and intensified difficulties for many economies.

What has come out clearly from the events of the past few years, however, is the relative ease with which the new level of oil prices has been integrated into the world economy. Despite fluctuations, prices of other commodities did not soar, and the world has been made aware of the scarcity of fossil fuels. The search has begun for alternative sources of energy, but the adjustment is hardly begun, let alone complete.

Development and Production in the Oil-Producing Countries

For the oil-producing countries, the past five years have been essentially experimental. Neither they nor the rest of the world knew whether or not the new levels of prices would be sustained over the medium term. Although they became conscious of their increased wealth, they were understandably unable either to gauge the dimensions of their increased income or to ascertain their new preferences. As far as oil production is concerned (and after the short-lived Arab embargo), they were not clear as to the volumes they should be producing at the new prices, but, since demand appeared to remain roughly the same as before, they experienced no significant change in export volumes. The countries with small petroleum reserves relative to annual extraction, however, have tended to value conservation (e.g., Libya) and the poorer and densely populated countries with growing demands to finance development (e.g., Indonesia, Nigeria, Algeria) aimed for maximum production. But there has been no scramble for markets owing to the balancing effect of these two tendencies, and to the fact that the market appeared accommodating to the total oil supplied by OPEC at the declared prices. In other words the prices charged were roughly right. Saudi Arabia, with its obvious interest in the longer term, tended to exercise a moderating influence on the course of short-term prices.[10]

With production thus fairly given, higher prices reflected themselves in higher incomes for all the oil producers.[11] It was fairly easy to adjust expenditure upwards, and this was done in the manner outlined earlier in the section of this paper on international payments through vast increases of imports. Expenditure on consumption rose considerably, particularly its component, public consumption, which includes defense and public administration. As mentioned before, the oil countries were also especially generous with economic aid, disbursing large amounts of resources bilaterally and through various funds and multinational agencies.[12] But the major portion of their expanded expenditure went to the much cherished goal of economic and social development, and practically every oil-producing country either began or expanded already begun medium-term development programs. With some exceptions, the states with large populations and other complementary resources had clearer visions about the future than the relatively empty desert economies whose goals were less clear. To the latter, which amassed the largest surpluses, development in the first instance often meant infrastructural development, and the bulk of the developmental effort was therefore directed at construction. Large imports of bulky construction materials and foreign labor became necessary and ports and other transportation facilities became congested, skills short, and inflation intense.

Infrastructure, however, is normally a support for productive activities and should be developed only to the extent that it is needed to serve these activities. In many instances the development of infrastructure has surpassed likely potential. With water and labor scarce, agricultural development proved to be unpromising and industry handicapped by lack of raw materials, labor, management, technology, and markets. Many oil producers with small populations therefore fell on the notion that their industrial development should be based on the processing of hydrocarbons

---

[10]See Appendix, "OPEC As a Cartel."

[11]Revenues would be a better word than incomes. Strictly speaking, proper calculation of income must take account of any asset depreciation. The pratice, so far adopted everywhere, to calculate GDP of the oil-producing countries does not allow for the depletion of their oil reserves. Should this depletion be charged as asset depreciation against the gross value added by oil extraction, very little income generated would remain. This is a fundamental conceptual point which is often ignored.

[12]According to the Organization for Economic Cooperation and Development (OECD), in the period 1973-1977 total official flows from OPEC members, on the basis of net disbursement, aggregated $31.5 billion and amounted to over 3 percent of their combined GNP. OECD, *Development Co-Operation*, 1978 Review, Report by the Chairman of the Development Assistance Committee (Paris: OECD, November 1978), p. 260.

into petrochemicals, especially where gas, currently being flared, could be utilized. A few energy-intensive projects in aluminum or steel are also being pursued. For all this, large-scale importation of labor and management has ensued, and the foreign influx, including workers' dependents, besides causing social problems, needs to be housed and fed, transported, educated, and given health care when ill; these tasks in turn create their own labor multiplier in that the imported labor necessary to perform them requires additional labor to provide more services for them and so on. Some countries therefore have decided to limit their future growth to reduce this foreign labor influx and also to moderate inflation.

During the past five years and in the flush of increased oil revenues, many countries lost sight of the need to economize, failed to coordinate development plans sufficiently with others in the same position, and ended up by bidding resources away from each other to the detriment of their terms of trade. In industrial development and even in infrastructure they have often duplicated facilities and created a potential for conflict over future marketing. With few exceptions it is fair to say that the major task of creating viable sources of income through economic and social development to replace petroleum remains as unattainable as ever. The past five years, in fact, have witnessed a waste of resources at a time when many parts of the world, with ample complementary resources, could have benefited from increased investments. At the heart of the waste lies the understandable but irrational desire to confine investments within individual national borders. While noting this, one must be aware nonetheless of the efforts of the Organization of Petroleum Exporting Countries (OPEC) and the Organization of the Arab Petroleum Exporting Countries (OAPEC) and of the various development funds in spreading investments throughout the Middle East and beyond. But it is the overambitious development plans and not these funds that are exhausting the bulk of the oil revenues, and it is these plans which need to be debated, rationalized, scaled down and coordinated in a regional context for the benefit of the oil producers primarily, but also for the benefit of their regions and the less-developed world generally.

From now on, as argued before, the pressure will mount on the oil producers with spare capacity to respond to the growing world requirements of fossil fuels by expanding their production particularly after 1985. And these very countries now may be having second thoughts about the waste that resulted from having produced too much oil during the past five years. As in Mexico, the desire may surface for these countries to tailor their future petroleum production to their ability to make effective use of the generated revenues.

Increasingly the oil-producing nations have come to realize their interdependence with the rest of the world. Their fortunes have become part and parcel of a world economy in which they now have an enlarged role. As producers of petroleum they require and receive technological advice and even management of their oil and gas fields which need continuous attention. Decisions on the rates of well production, secondary recovery, and use of associated gas are among many that have to be taken in consultation with their production partners and marketers. The bulk of the market for their valuable product lies outside their boundaries. The freedom of the seas should be as much their concern as that of the oil consumers. Shaping and implementing their development plans often depend on outside support. Even their consumption needs have now to be satisfied by imports, and their financial investments will forever depend on the potential and economic stability, prosperity, and sanctity of contract in the various countries in which they have already invested and are likely to continue to invest in the future.

There is little doubt in this writer's mind that the oil-producing states should respond positively if required by the oil consumers to increase production, for it is not in their interest to impede growth in the industrialized countries or inflict hardships on the poorer and less-developed nations. Expanded production, however, would mean reduced prices (reduced, that is, compared with the prices that would otherwise prevail), but it is not beyond the ingenuity of man to devise a method of fair compensation to the producers for any loss of revenue due to expanded production. But apart from price, which is a lesser issue, the producers should not regard excessive financial resources as inevitably wasteful. The world outside is hungry for investment funds, and again it should not be beyond the rationality of man to devise ways and means to channel oil surplus funds into productive investments anywhere in the world: to increase food and energy and employment-creating industrial production.[13]

The world wasted a great deal of resources and time through the exaggeration of the problem of the balance of international payments, but in retrospect we know that that was an illusory problem. Waste of resources through public and private expenditures in the newly rich oil countries was tolerated, and in some instances even encouraged, to boost their imports. In retrospect it is easy to see that we are all the poorer for this waste. The lesson of the past few years seems clear. There should be a concerted effort for cooperation among all parties. The consumers want an assurance of

[13]In this context, one should note the recent decision by the World Bank to embark on an energy development program in less-developed countries. Increased production of oil and gas as a result of this program may not add substantially to the supply of petroleum in the world market and will take some time to materialize, but it is a step in the right direction and should alleviate some of the hardship in a number of the poorer countries. Additionally , it should contribute to lessening the pressure on oil producers to increase sales.

security of supply and some elasticity over time in response to growing demands. The producers want a fair price that reflects the growing scarcity of their exhaustible resource. If sinking excessive resources in the producers' own economies is wasteful, such resources should be invested in financial or physical assets that have to lie outside the oil producer countries themselves. Much dispute over oil prices and the volume of oil output has come from the fact that no effective outlets have been developed to afford the producers equitable returns on their investments. Such returns, as noted earlier, at equilibrium should be equal to the returns the producers can legitimately expect from leaving their oil in the ground to appreciate. The fact that the producers have been paid in depreciating dollars, and that their financial investments have often carried negative real returns, is not encouraging for production expansion should this be needed in coming years. For the future it should be possible for oil to be priced in a currency other than dollars, say in the International Monetary Fund's Special Drawing Rights (SDRs) and investments could likewise be denominated also in SDRs carrying positive interest rates. Side by side with the raising of returns on financial investments, returns on the physical formation of assets also should be boosted. If the oil countries have no immediate investment opportunities within their borders, they should expand their outside investments. The major development funds should be integrated and their resources greatly expanded. The primary target should be to promote regional development, and after this to promote development anywhere else in the less-developed world.

Regionalism in the Middle East may be viewed as comprised of the Arab League. It may come as a shock that if all the income of the region including that of Saudi Arabia, Kuwait, and the United Arab Emirates is evenly divided among the population of the countries of the Arab League, income per capita would amount to no more than $1,134 in 1976, as compared with a world overall average of $1,697. The Arab world average income, therefore, is only about two-thirds of that of the world average (see tables 3 and 4 of the Appendix for details). In other words, the Arab countries, taken as a whole, are not a rich part of the world — except perhaps in potential, and for the potential to be realized a great deal of hard work and vision is necessary, and the help and cooperation of the outside world is essential.

Conclusion

To sum up, it may be helpful to emphasize a few points elaborated in this paper. The world has become so interdependent that it is no longer possible for any one group to realize significant gains at the expense of others without damaging its own interests. The outside world has so far failed to develop substitutes for petroleum, and the higher prices of 1973-1974 have more or less prevailed; but the search for alternatives has been put in motion, and it is just a matter of time before these substitutes are developed. The producers therefore must attempt to develop other sources of income. The oil exporters already have considerable financial interests through investments in other parts of the world and thus have become much more sensitive than before to peace and prosperity globally. World requirements of petroleum are projected to grow at the current level of real prices and if the present conditions should prevail, the oil producers will lack sufficient incentive to expand production and ease the pressure on prices to rise. The prevailing conditions include payment for oil in depreciating dollars, negative-bearing financial investments, and wasteful expenditure at home. To induce the oil producers to maintain and even expand their oil output there should be cooperation by all sides to offer them a fair price for their product, boost their financial returns through inflation-proof investments which are safe from exchange depreciation, and offer them the opportunity to channel their investment outside their borders where the returns on investments should be higher. If this is done, they are bound to respond by offering the world the energy supplies it will so badly need.

Appendix

On OPEC As a Cartel

The Organization of the Petroleum Exporting Countries (OPEC), which began as a consultative forum for the oil-producing countries in 1960, is now often referred to as a cartel. A cartel seeks to raise the price of the cartelized product over and above the price that would rule if free competition prevailed. The logic behind a cartel is that the conditions of demand for the product are such (e.g., price elasticity being lower than one) that a smaller supply would fetch higher revenue than a larger supply. A cartel would seek to limit supply and allocate the limited supply as quotas among its members (usually by reference to some historical sales). In commodity markets there have been many instances of cartels being formed only to break up after a short period of time. The artificially higher prices they maintain bring forth supply expansion outside the cartel, and also induce buyers to seek substitutes for the product; the market would then revert to competition.

Although the members of OPEC had consulted with each other about policy prior to 1973, the characterization of OPEC as a cartel did not become current until the oil price revolution of 1973-1974. The fact that OPEC began afterwards to hold periodic conferences, at which a price was declared for OPEC members, reinforced the image of the organization as a cartel. It was immaterial whether or not OPEC limited supply. By fixing price, total supply was determined. Although there was no allocation of quotas among the various producers, Saudi Arabia appeared to play a balancing role of adjusting its own production so that the declared oil price could prevail. Saudi Arabia seemed to be in the special position of being fairly indifferent to the level of its production since its financial requirements could be accommodated within a wide range of possible levels of production. Saudi Arabia thus appeared to perform the necessary limitation of supply on which a cartel has to depend.

Little attention was given to the fact that the $2 per barrel price that preceded the price revolution of 1973-1974 had been determined in a market in which the oil multinationals enjoyed great powers (supported often by the military and political powers of their principals) over drilling, the rate of extraction, technology, distribution, and prices. That they acted often as oligopsonists with influence on prices and revenue-sharing with the host countries cannot be denied. Gradually, however, the powers of the host countries did increase *pari passu* with the decline of the old empires and the changed balance of world power.

The abrupt rise in prices under the aegis of OPEC in 1973-1974 caused passionate reactions. As argued in the text of this paper, this rise was only possible because the price had been depressed for a long time, and the world had become addicted to cheap petroleum as a major source of energy. Many analysts, some of them eminent, believed that the petroleum market had been competitive before and was now being artificially cartelized. There was also large-scale confusion over the cost of extraction which was often identified with the cost of production. Account tended not to be taken of the fact that petroleum was an exhaustible resource.

It was therefore common to believe in 1974 and 1975 that the oil "cartel" would eventually have to disband, and the artificially high prices would collapse as the industry reverted to its previous state of "competition." Professor Houthakker wrote:

> There may be markets where the price can be doubled or tripled without changing the balance of supply and demand, but the petroleum market is certainly not one of them. The incentives for increasing production are tremendous, and no matter what the Club of Rome may say, the potential for greater output exists, not only in the currently exporting countries, but also in the United States. In the very short run not much can be done, but at prevailing OPEC prices we are likely to see a surplus of oil within the next year or two. Normal responses to higher prices on the demand side will work in the same direction. The producer cartel will then have to adopt prorationing if it wants to keep the price up.[1]

Fundamentally, people who took the Houthakker line, and they were legion, believed that the new level of prices was sustainable only by supply limitation, and that OPEC was truly a cartel.[2] If, on the other hand, the higher prices are judged to be more in harmony with competitive conditions, being sustained essentially by fundamental market forces, then the concept of a cartel becomes superfluous, and it is the pre-1973 market form that was then at variance with competition. In the present conditions of oil shortage, with Iran's production interrupted, it is easy to defend the latter position. The OPEC declared prices now are being exceeded in bids in the spot market (early 1979), and individual producers are raising their

---

[1]H.S. Houthakker, "Policy Issues in the International Economy of the 1970s, "*American Economic Review, Papers and Proceedings*, May 1974, p. 140.

[2]It was a strange cartel whose members behaved discordantly at the height of the Arab embargo in 1973. The non-Arab members of OPEC did in fact expand production to take advantage of the reduction of Arab supply, and some of the Arab members themselves disagreed with the embargo. OPEC also has been unable to regulate the premia and discounts (for quality and location) around the Saudi marker crude, and these have often fluctuated widely. Moreover, some OPEC members resorted to bartering their oil at prices which could not be known.

prices without consultation with each other. But even if Iranian production is fully restored, the price of oil is expected by a growing number of analysts to increase not to fall over time. The world has had five years to adjust to the higher prices, but substitutes on either the demand or supply side of the market have been scarce. In these conditions if OPEC were to be liquidated voluntarily, the price of oil would probably not fall. Thus OPEC does not seem to be the cartel it is held to be.[3]

In the recent past, particularly since the Doha meeting of OPEC in December 1976 when a two-tier price reflected internal dissensions, it has become obvious that the organization is unable or unwilling to achieve a consensus that is truly binding on its members as to price and, therefore, volume of production. Sporadic increases in prices, declared by individual members of OPEC in early 1979 in reaction to the Iranian crisis, also indicate that OPEC's real function has become one of providing a forum for the oil exporters to consult with each other. Such consultations are necessary and useful, but it is high time that they were extended to include the major oil consumers. OPEC would be an ideal venue for such consultations. Even the acronym, OPEC, could remain unchanged, standing for the Organization of Petroleum Exporters and Consumers, and the OPEC would henceforth be rid of the cartel image it has unfairly assumed.

Table 1

OIL PRICE TREND
(U.S. dollars per barrel)

Year	Saudi Arabian Realized Price[a] (current)	Index of Price of Manufactures[b] (1975=100)	Saudi Arabia Realized Price Deflated by Index of Price of Manufactures (1975=100)
1950	1.71	36.5	4.70
1955	1.93	42.3	4.56
1960	1.50	45.0	3.35
1965	1.33	46.6	2.87
1970	1.30	52.9	2.46
1975	10.72	100.0	10.72

Source: The World Bank, *Commodity Trade and Price Trends*, 1977 Edition.

[a]Light crude oil, 34°-34.9° API gravity, f.o.b. Ras Tanura.
[b]Prices of manufactures (SITC 5-8) c.i.f. index converted from an index based on 1970. Classification includes chemicals, manufactured goods classified by materials, machinery and transport equipment, and miscellaneous manufactured articles.

---

[3]Practically all oil producers, with the exception of Saudi Arabia, produce up to the limits of their capacity — capacity being defined to take depletion into account, not just the transient facilities that happen to be in place. Saudi Arabia could, of course, deluge the world with oversupply, but this would be contrary to its own interests as it lacks alternatives for financial investments that would give it returns equal to the returns that would accrue through appreciation by leaving its oil in the ground.

Table 2

OIL-EXPORTING COUNTRIES:
EXPORTS, IMPORTS, AND INTERNATIONAL RESERVES[a]
(in billion U.S. dollars)

	1972	1973	1974	1975	1976	1977	1978
Exports	24.27	38.34	118.46	109.42	132.88	145.30	(67.01)[b]
Imports	13.50	19.54	32.41	51.28	62.61	85.12	(48.86)[b]
International reserves (total end-year)	10.90	14.35	46.16	58.63	64.82	72.42	58.38[c]
International reserves (addition during year)	3.10	3.45	31.81	12.47	6.19	7.60	-14.04[d]

Source: International Monetary Fund, *International Financial Statistics.*

[a]Cover Algeria, Indonesia, Iran, Iraq, Kuwait, Libya, Nigeria, Saudi Arabia, United Arab Emirates, and Venezuela.

[b]First half of 1978 only.

[c]At end of November 1978.

[d]Eleven months until the end of November 1978.

Table 3

POPULATION, GNP PER CAPITA AND TOTAL GNP OF MEMBERS OF THE ARAB LEAGUE, 1976

Country	Population, Mid-1976 (000)	GNP at Market Prices 1976 US$ (million)	Per Capita (US$)
Iraq	11,510	15,960	1,390
Saudi Arabia	9,240	40,860	4,420
Syrian Arab Republic	7,655	6,320	830
Yemen Arab Republic	5,406	1,620	300
Lebanon	3,266	n.a.	n.a.
Jordan	2,792	1,820	650
Yemen, People's Democratic Republic	1,743	470	270
Kuwait	1,030	14,380	13,960
Oman	796	2,090	2,620
United Arab Emirates	694	10,050	14,480
Bahrain	320	1,220	3,810
Qatar	210	2,440	11,460
Egypt, Arab Republic of	38,228	10,680	280
Morocco	17,197	8,900	520
Algeria	16,463	16,700	1,010
Sudan	16,127	4,390	270
Tunisia	5,732	4,580	800
Somalia	3,579	400	110
Libya	2,537	15,140	5,970
Mauritania	1,495	380	250
Djibouti	273	160	580
Total Arab League[a]	139,761	158,560	1,134

Source: The World Bank, *1978 World Bank Atlas*.

[a]Excluding Lebanon and "Palestine."

Table 4

WORLD POPULATION, GNP PER CAPITA AND TOTAL GNP BY MAJOR REGIONS, 1976

Region or Country[a]	GNP Per Capita 1976 (US$)	GNP 1976 (US$ billion)	Population, Mid-1976 (millions)
North America	7,880	1,877	238
Japan	5,090	574	113
Oceania	5,320	115	22
Europe, excluding U.S.S.R.	4,280	2,215	518
U.S.S.R.	2,800	718	257
Middle East	2,250	176	78
South America	1,230	270	219
Central America	1,000	109	109
Africa	420	180	426
Asia, excluding Japan and Middle East	290	586	2,040
World	1,697	6,820	4,020

Source: The World Bank, *1978 World Bank Atlas.*

[a]For definitions of area refer to source.

8

OPEC AND THE PRICE OF OIL: CARTELIZATION OR ALTERATION OF PROPERTY RIGHTS

*Ali D. Johany**

It is widely believed that the current price of crude oil reflects the monopoly power of OPEC rather than the long-run supply costs of its production. The successful cartelization which is being attributed to the OPEC organization has become such an intrinsic part of current economic (let alone popular) literature and so widely accepted that even to entertain the doubt that such is not the case may shock most readers. Nevertheless, in this paper it will be argued that the price of oil would have sharply risen in January 1974, independent of the presence or absence of OPEC, given that the individual oil-producing countries became the ones who determine the rates of oil output at different market prices at the end of 1973.

The Cartel Question

Toward the end of 1973, everyone was aware that the price of oil rose sharply but few knew (or fully appreciated) that a *de facto* nationalization of the companies' crude also took place. Due to the apparent suddenness and the magnitude of the price change, many economists (including the writer) had concluded that OPEC became an effective cartel that started to exercise its monopoly power. Furthermore, economists thought that, like other cartels, this one would soon crumble. In the words of Professor M.A. Adelman, one of the foremost authorities on the petroleum market: "Unless the producing nations can set production quotas and, what is more impor-

*Ali D. Johany, presently Dean of the College of Industrial Management, University of Petroleum and Minerals, Dhahran, Saudi Arabia, earned a Ph.D. in economics from the University of California, Santa Barbara and earlier served with the Saudi Arabian Central Planning Organization. The author prepares a weekly commentary on economics-related topics for *Al-Riyadh* newspaper and is currently involved in a study on world energy and Saudi Arabia. Dr. Johany wishes to acknowledge the useful comments and suggestions made by Professors Walter Mead, H. E. Frech III, and Robert Deacon.

tant, obey them, they will inevitably chisel and bring prices down by selling incremental amounts at discount prices."[1] After he explains the incentive for cheating, he concludes: "The World Oil Cartel of the 1930s was eroded by this kind of competition, and so will be the new one [OPEC] in the 1970s."[2]

As a consequence of the durability of the alleged OPEC cartel, Adelman changed his mind. His explanation for the apparent robustness of OPEC is based on an institutional factor which he believes to be peculiar to the member countries of the organization. According to Professor Adelman, when OPEC sets its price at a certain level it does so by raising its member governments' taxes per barrel of oil.[3] These taxes that have to be paid by the oil companies "are in the form of income taxes, in fact excise taxes, in cents per barrel. Like any other excise tax, they are treated as a cost and become a floor to price."[4] If it were not for this OPEC tax system, the "cartel would crumble." This explanation is not sufficient to explain the durability of OPEC because, analytically, there is no meaningful difference for an oil-producing nation between reducing its own price or reducing its marginal tax rate to increase its market share.

The Alteration of Property Rights

In the last three months of 1973 there was an oil embargo; there were output reductions; but something else happened, perhaps more important than all of these things. On 16 October 1973 the oil producers decided to determine the price of their oil unilaterally rather than through negotiations with the oil companies as had been done in the past.[5] Once the host countries

---

[1]M. A. Adelman, *The World Petroleum Market* (Baltimore: The Johns Hopkins University Press, 1972), p. 258.

[2]Ibid.

[3]M. A. Adelman, "Is the Oil Shortage Real? Oil Companies as OPEC Tax Collectors," *Foreign Policy*, winter 1972-73, pp. 69-107.

[4]Ibid., p. 76.

[5]Because of the absence of a central oil market, it is difficult to know the precise price of a unit of oil at any time. The "posted" price was employed as an internal accounting device that the oil companies used to transfer oil from their producing to their refining and marketing subsidiaries. Prior to 1973, the posted price was also used as a basis for the companies' payments to the producing countries and was determined by the bargaining power of the companies on one hand versus the host countries' governments on the other, and thus might have been radically different from the market-determined price.

became the ones who decide how much to produce and how much to charge for each barrel of oil, the role of the companies was essentially reduced to that of contractors. That amounted to a *de facto* nationalization of the crude oil deposits. The crux of this paper is that this *de facto* nationalization is what caused the substantial change in the world oil price rather than the alleged successful cartelization of OPEC.

Prior to World War II, the oil-producing countries of Latin America and the Middle East granted some of the major Western oil companies the rights to search for and produce oil. The terms of the original concessions dictated that the producing nations receive "royalty" per unit of output and gave the companies the right of ownership of crude oil deposits. These agreements were repeatedly renegotiated. Each successive agreement resulted in better terms in favor of the countries. Gradually the property rights of the companies in crude oil were eroded until in 1973 the companies became mere production and marketing agencies instead of owners of crude deposits. The forces that increased the economic and political power of the countries and decreased those of the companies will not be investigated here.[6] For the purpose at hand, the essential point worth emphasis is this: property rights in crude oil deposits had changed from the companies to the countries. Furthermore, the oil companies were aware of the gradual shift of ownership.[7]

Since the sharp increase in the market price of oil that followed the October 1973 Arab-Israeli War (although it is not certain that prices were quadrupled as has been frequently asserted), economists have assumed that OPEC has become an effective cartel that reduces output to raise prices. This writer disagrees with such an explanation. There is no doubt, of course, that without production cutbacks prices will not rise. The only disagreement is with the reasons which were advanced to explain the production restrictions.

The Influence of Actual and Expected Nationalization on Rates of Output

The objective of an owner of oil deposits, as with any owner of any other asset, is to maximize its discounted present value. Thus, if he expects his deposit to be exhausted (in the sense that the demand for it will fall to zero) in period T, then he wants to choose the time path of outputs in such a manner as to maximize the discounted present value of the deposit. Let us define

---

[6]See Ali D. Johany, "OPEC is Not a Cartel: A Property Rights Explanation of the Rise in Crude Oil Prices" (Ph.D. dissertation, University of California, Santa Barbara, 1978), chapter 5.

[7]J. E. Akins, "The Oil Crisis: This Time the Wolf Is Here," *Foreign Affairs*, April 1973, pp. 463-90.

the net value of a unit of oil R_t for any period t as

$$R_t = P_t - C_t$$

where P_t and C_t are respectively the unit price of oil and the average cost of producing it in period t. For simplicity, assume constant costs, $\frac{dc}{dt} = 0$. For an oil company to supply a positive amount of oil in any two periods t and t + 1, the net value of a unit of oil must rise over time at a rate equal to the market rate of interest (r). That is, under a world of certain and well-defined property rights, an oil company will supply oil in the two periods t and t + 1 if it expects

$$R_{t+1} = r_t (1 + r).$$

The oil companies in the pre-1973 era, however, knew that some threat to their property rights in the OPEC countries always existed. The probability of expropriation varied among nations, but the companies were never 100 percent certain that their property rights would not be threatened; and that had great influence on their behavior.

Let us say there is a probability $q<1$ that nothing will interfere with their operations.[8] Then,

$$E(R_{t+1}) = qR_{t+1} + (1-q)0$$

$$\text{or } R_{t+1} = qR_{t+1}. \qquad (1)$$

Thus, as a consequence of any chance of nationalization, the companies will supply oil in period t and period t + 1 only if

$$R_{t+1} = R_t (1 + r), \qquad (2)$$

that is, R_{t+1} must be greater than $R_t (1 + r)$ in the case of uncertainty. From (1) and (2) one can see that positive amounts of oil will be supplied by the companies in the two periods if

$$qR_{t+1} = R_t (1 + r)$$

$$\text{or } R_t + 1 = R_t \frac{(1 + r)}{q}.$$

[8]The assumption of either nationalization with no compensation or no threat to property rights of any kind was made for simplicity. The conclusion will not differ if we assumed the more probable case where some token compensation is made or when the governments erode the property rights of the companies by reducing the concession areas or by increasing the companies cost of production by imposing additional excise taxes.

The effect of less than 100 percent certainty about future property rights is to increase the effective rate of discounts, r, on which the oil companies will base their decisions. For example, say the market rate of interest, r = 10 percent, and say a company thinks that the probability that no one will interfere with its operations is 0.75. Then

$$R_{t+1} = R_t \frac{(1+0.1)}{0.75}.$$

Therefore, $1 + r = \frac{1.1}{.75} = 1.47$, or, the companies' effective discount rate is 47 percent (!) when r = 10 percent and q = 0.75. The net effect of uncertainty of property rights is to increase the companies' discount rate which will lead then to increase their oil outputs by a greater rate than they otherwise would if there were no risk of expropriations. Figure 1 shows two possible paths of (P-C), given r > r. The horizontal axis measures the life of a deposit; the beginning of production is period zero and the time of exhaustion is period T. The vertical axis measures the net value, (P-C).

Figure 1

THE INFLUENCE OF INCREASING DISCOUNT RATES
ON PRICES OVER TIME AND ON DATE OF EXHAUSTION

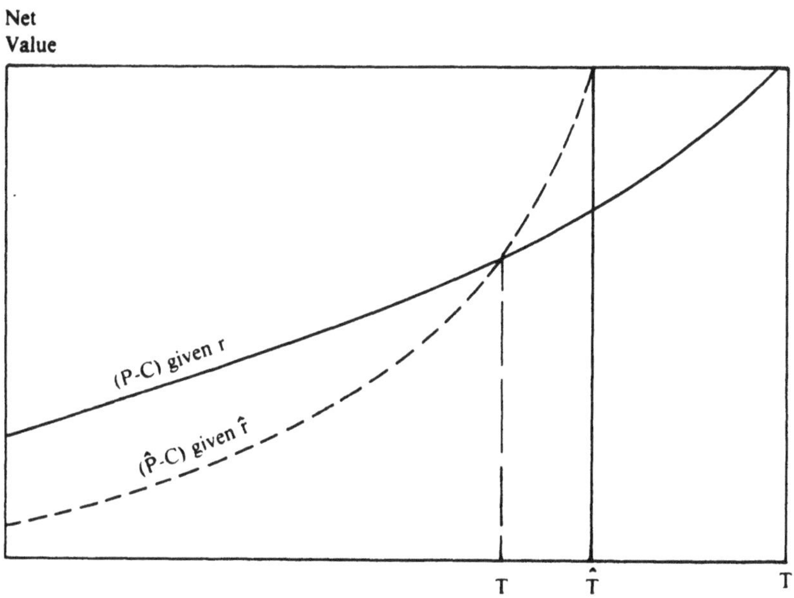

Economic Decisions of the Oil Companies versus the Host Countries

From the above analysis we can deduce that the presence of uncertainty will make the supply of oil at anytime between periods 0 and T greater than what it will otherwise be given certain and secure property rights. Conversely, the supply will be smaller in the periods between T and T. Of course, it is to be expected that uncertainty will result in greater rates of output. What is surprising is the fact that at $q = 0.75$, a reasonable approximation given the political instability of the Middle East, the effective discount rate has changed from 10 percent almost fivefold to 47 percent.[9] It must be the case, then, that prior to October 16, 1973, when the private Western oil firms lost their claims on crude oil, the supply of oil was artificially large since the companies had learned from previous events that the day of nationalization was forthcoming. An oil company may have been aware that greater outputs meant lower prices. Yet, history must have taught the executives of oil companies that a barrel of oil left in the ground today may either never be available or only available at a much higher tax rate.

The host countries' governments, on the other hand, faced no threat of nationalization, and as a result will produce oil now and in the future as long as they expect the net value to rise over time at a rate equal to the world rate of interest.

The fear of nationalization, which led the firms to act as if "there were no tomorrow," is the main reason for the difference in behavior between the foreign companies and the host nations. But that is not all. Even if the companies faced no threat to their property rights whatsoever, they would have produced more oil than the governments. One reason is that most of the original concession agreements, giving the firms exclusive crude ownership rights, would have expired by the 1990s. Another reason is that most of the oil-producing countries do not have the same investment opportunities that the companies have.

As a consequence of the present stage of economic development in many of the OPEC nations and as a result of the relative abundance of financial assets in these countries, there are only a limited number of projects that could be undertaken before the rate of return on domestic investment falls below the world market rate of interest. If, on the other hand, the OPEC members wanted to invest some of their funds abroad, then they are uncertain that their property rights in financial assets will not be threatened. That is, the host countries have a problem of insecure property rights in their investment in foreign nations which is qualitatively similar to the problem faced by the oil companies in the OPEC states.

---

[9]The value assigned to q is necessarily arbitrary. As long as $q < 1$, and it must be, the effective rate of discounting will be greater than the market rate of interest.

The price of oil has risen sharply since 1973, but the pre-1973 price level might have prevailed due to the artificially large oil output that the companies' economic constraints forced them to produce. In the 1950-1972 period, the average annual gross additions to the free world's oil reserves were roughly 27 billion barrels.[10] That might have seemed to mean the oil companies had a choice between increasing output, and thus reducing prices, or keeping prices at a higher level by leaving a lot of oil in the ground when their concession terms had expired or when nationalization had occurred. An individual firm, however, could not have much of a choice because the oil market gradually became competitive, and any single producer's efforts to prevent a price fall by output cutbacks would have resulted in the reduction of its own revenues. Since October 16, 1973, increases in oil reserves ceased to mean automatic increases in supply.

In other words, the post-October 1973 oil prices are much higher than the pre-October 1973 prices mainly because the earlier period's prices reflected the supply costs as viewed by the oil companies; the latter period's prices reflected the supply costs as viewed by the oil-producing countries. To say that the cost of producing a barrel of oil in the Middle East is between 20 and 30 cents does not explain much. Since the oil-producing nations need not fear the loss of their property rights (excluding foreign invasion) or the expiration of the concession terms, their decision as to how much oil to produce in each period is influenced by their discount rates (which are much lower than those of the companies) and the expected future prices, as well as by the current and future costs of exploitation. Given the limited investment opportunities in most of the producing countries, one could not conclude correctly that their output rates are inconsistent with wealth maximization behavior, aside from any effort on their part to keep the "cartel" together.

The large differences between the oil prices that prevailed before and after the end of 1973 and the slight restriction of oil output that accompanied the price changes do not contradict the monopoly power that is being attributed to the successful alleged cartelization of OPEC. What contradicts that explanation is the absence of the effective mechanism which is needed to keep the cartel going. Besides the political differences among the OPEC countries — they vary from radical socialists to conservative monarchies — the economic differences (such as size of oil reserves, size of population, stage of economic development, and the like) alone will make agreement on prices and rates of output very difficult.

Disagreements on prices and production rates could be solved, however, if side payments and demand prorationing were effected. But no such things occur. In the early 1960s, long before OPEC was thought of as an effective

[10] J. H. Lichtblau, "Factors Shaping Future Petroleum Prices," *Oil and Gas Journal*, August 1977, pp. 510.

cartel, all attempts within the organization toward production quotas and demand prorationing were abandoned. A satisfactory formula for output quotas could not be found.[11] Venezuela, then the largest producer, wanted historical levels of output to be used as the base, Iran argued for national population, and Saudi Arabia and Kuwait selected proven reserves as the base for production quotas.

The fact that the oil-producing governments' representatives meet to "fix" prices is not peculiar to OPEC. The producers of coffee, tin, copper, and many other raw materials meet frequently to fix the price of their commodities. But unfortunately for colluders, as George Stigler has convincingly argued, it takes far more than price agreements to effectuate and maintain profitable collusion.[12]

Conclusion

Cartels — legal, illegal or whether their members are sovereign states or private firms — that are able to earn monopoly profits for their members are inherently unstable. The price of oil did not fall substantially since its rise on October 1973, not because OPEC defied our economic theory of cartels but because OPEC is not a cartel. In this brief paper an attempt has been made to show that the constant threats to the oil companies' claim on crude oil and the limited investment opportunities that are available to some of the OPEC countries caused the discount rates of the companies to be much higher than those of the host countries. And different discount rates lead to different rates of output, which in turn lead to different prices.

The rise in prices not only increased the OPEC producers' revenues but also decreased their discount rates even further. That is because the additional increase in revenues was not accompanied simultaneously by an increase in domestic productive capacity (which allows increasing investment expenditure without decreasing real rates of return) and because investment abroad, as an alternative to domestic investment, may be subject to expropriation. Therefore, the increase in revenues, through its influence on discount rates, dictated further output reductions which in turn helped prevent the oil price from falling.

From the late 1940s to the early 1970s, the real price of oil had fallen by about 65 percent.[13] The price of oil was then, and still is, being determind by

[11] This has been traced in Z. Mikdashi, *The Community of Oil Exporting Countries* (Ithaca: Cornell University Press, 1972).

[12] G. J. Stigler, *The Organization of Industry* (Homewood, Illinois: Richard D. Irwin, Inc., 1968), chapter 5.

[13] M. A. Adelman, *The World Petroleum Market*.

demand, long-run supply costs, and the extent of competition in the oil market. I have argued that what has changed is not, as has been frequently asserted and widely accepted, the elimination of competition, but rather the long-run costs of supplying oil. Since the alteration of property rights, increases in reserves ceased to imply automatic increases in supply. As far as any oil-producing nation is concerned, the time of exhaustion is not influenced by threats of nationalization or the expropriation of the concession terms. Thus, the pre-October 1973 oil price reflected the true long-run costs of production as viewed by the Western oil companies, and the current price may reflect the actual long-run production costs as viewed by the oil-producing nations.

In real terms the price of crude oil has been falling since its sharp rise in 1974 because Saudi Arabia was always willing to increase its rate of output to hold down further increases in the world oil price. And if the consuming Western nations were to enact laws that somehow give the producing countries greater certainty on their investment activities in the West, the price of oil would even fall further. The reason? The effect of such legislation would be to increase some of the OPEC countries' average rates of discount (because now higher rates of return could be realized on secure foreign investment) and that would induce them to produce more oil in each period of time.

9

FUTURE PRODUCTION AND MARKETING DECISIONS OF OPEC NATIONS

*Dennis J. O'Brien**

OPEC and the other petroleum-exporting nations hold most of the cards in world crude oil trade in the 1980s. With the operational aspects of the industry increasingly under their control since 1973, the oil-exporting countries can vary current production and future capacity levels to meet their social, economic, and political needs.

Increased marketing activity by producer national oil companies since 1979 has added another dimension to the power and influence these entities hold on world oil markets. We do not know how far or how fast these companies will go in their desire to become integrated enterprises, both inside and outside their national boundaries, but we can be sure they have a number of options in the production and marketing areas of the industry as well as in those areas that link the two. This growing operating and marketing control, combined with the gradual exhaustion of accessible deposits of conventional world crude oil, will cause oil production to plateau or perhaps decline in the next few years. If political events or natural catastrophies occur, the decline in oil production could be considerable.

New or additional discoveries in promising new areas – offshore North America, China, Sudan – could add to supply by the mid- to late 1980s, but would not offset the decline of existing fields. The same can be said about increased exploration within OPEC and other major suppliers: success can delay but not offset the expected decline in the '80s.

It is likely that output in the Arabian/Persian Gulf countries will at best remain near current levels; production in other OPEC nations will decline;

* Dennis J. O'Brien, now with the International Affairs Division of the United States Department of Energy, was previously a Faculty Fellow at the General Accounting Office, on leave from his professorship with the International Center, International Affairs Graduate Program of California State University at Sacramento. The author received his Ph.D. from the University of Missouri, specializing in foreign policy, international affairs, energy and resources, and diplomatic history. The views expressed in this article are those of Dr. O'Brien only.

the centrally planned economy bloc will cease to be net exporters and become net importers; and OECD nations' oil production will steadily decline after 1985. Some increased supply from the non-OPEC less-developed countries (LDCs) could be available if exploration and development goes well in the new provinces noted earlier and if Mexico elects to aim at higher production targets.

Options

Let us turn to the major producers both in and out of OPEC. The first observation to be made is that some have options on the production side and others do not. Resource constraints, technical factors, and political problems narrow some nations' abilities to develop or maintain productive capacity, while other nations have a wide range of options in developing future capacity and output. Some OPEC leaders recently have been quoted as saying OPEC will have the capacity to produce 40 million barrels daily by 1985, but expects to produce only 30 million barrels per day (b/d). They are probably talking 3.5 million b/d sustained capacity. I would like to identify some areas and speculate what might be done with the 5 or so million b/d spare capacity.

On the marketing side, all exporting nations have options on how they might market crude. These range from simple processing agreements and destination controls to downstream integration into refining and transportation through equity or joint ventures. It is my belief that oil-producing nations in the next 10 years will integrate downstream for a multitude of reasons – many of them noneconomic.

Saudi Arabia

There is no question about Saudi Arabia's potential capacity to produce oil. The reserve figure typically used is about 163 billion barrels, which is the proved and probable reserves of the 15 Aramco producing fields. The Neutral Zone (jointly administered with Kuwait) would add another 3 to 4 billion barrels, while the non-producing Aramco fields are estimated to have proved and probable reserves of at least 15 billion barrels. Adding all of these together results in a total proved and probable reserve figure of about 182 billion barrels.

Saudi crude production, that is, Aramco production excluding the Neutral Zone, was 9.3 million b/d in 1979, and has been estimated to run at about 9 million b/d in 1980. Given its vast reserves, future Saudi production is primarily an investment and political decision. For its own foreign exchange needs, Saudi Arabia has to produce less than 5 million b/d. However, Saudi Arabia has a deep concern for the political stability of the West and the stability of international financial markets. Offsetting this interest to a significant degree is domestic opposition to high production levels from conservationists and technocrats. It is unclear what level of future production these factors will combine to yield. The safest estimate, however, is 8.5 million b/d in 1985 and 1990, which is the Saudi current desired maximum level of output.

The key to Saudi production lies in creative investment packages and increasing downstream involvement in refining and marketing. Recently, the Chairman of the Saudi Investment Corporation called for investment schemes that would provide higher long-term yields than anticipated oil price increases would as an incentive. Oil Minister Yamani is reported to have indicated that transfer of technology agreements may be necessary to renew oil supply contracts. Joint ventures with Mobil and Japanese firms in refining oil and producing petrochemicals give the Saudi government a stake in maintaining supplies and optimum operation of the refineries. Petromin's processing agreements and direct crude sales – particularly to local independent distributors outside Saudi Arabia and to other governments – also commit the Saudis to long-term agreements. Petromin's sales may top 2 million b/d by the end of 1980. The success of these initiatives will determine how much further Petromin will go and how much it will likely influence future production levels.

Kuwait

Kuwait, like Saudi Arabia, has the reserves (65.4 billion barrels) and resource potential to support oil production at levels higher than those now in place. Again like Saudi Arabia, Kuwait has no economic incentive to expand output; it, too, is a traditional state in a region of the world where an increasing number of radical leaders are coming to power.

The Kuwaitis could be very flexible on future production levels. Output could be expanded considerably based on their relatively enormous reserves. However, Kuwait currently needs to produce only about 1 million b/d to obtain its foreign exchange requirements. In order to supply local power generation plants and meet its existing liquefied petroleum gas (LPG) export commitments, Kuwaiti crude production (including the Neutral Zone) cannot fall much below 1.5 million b/d. This is my estimated production level for 1980.

Because of this flexibility, future Kuwaiti production will depend almost entirely on political and industrial factors. The country clearly wants to conserve oil and does not need to produce more in order to obtain sufficient foreign exchange. If a conservationist strategy is followed, output of crude oil would probably remain around 1.5 to 2 million b/d in 1985 and 1990.

Kuwait has replaced National Iranian Oil Company (NIOC) as the most aggressive national oil company in expanding downstream operations. The restructuring of crude supply contracts, including non-competition, no resale, destination clauses, and overall scrutiny of crude movements, is a strong indication that Kuwait is in marketing to stay. A recent Kuwait National Petroleum Company (KNPC) deal with Taiwan involving four product tankers costing $100 million points to an expansion into Eastern hemisphere marketing activity. KNPC also has expressed some interest in exploration and production outside the Arabian Gulf, and Kuwait's petrochemical industries recently have set up a joint phosphate venture in Tunisia. Where Kuwait goes, other OPEC national companies are likely to follow.

United Arab Emirates

Future crude production in the United Arab Emirates (UAE) will depend on the actions of Abu Dhabi and Dubai, its major oil-producing sheikhdoms. Dubai is expected to continue to produce close to capacity. It has limited reserves and is unlikely to expand output much above its current level of 300,000 b/d. The production of Abu Dhabi will depend on a number of factors. Abu Dhabi needs to produce only about 1 to 1.2 million b/d to satisfy its foreign exchange requirements. However, to keep its LPG plant operating at capacity, about 1.95 million b/d of output would be necessary by the mid-1980s.

The actual level of output, therefore, will depend largely on political rather than economic factors. The primary issue here will be the Arab-Israeli dispute and Abu Dhabi's concern over the economic health and political stability of the West. My best estimate is that Abu Dhabi production will be between 1.4 and 2 million b/d – probably at the upper end of the range. This implies UAE output of 1.7 to 2.3 million b/d in 1985 and 1990.

The UAE has not indicated a strong interest in downstream involvement in marketing as yet. A major refining expansion at Umm Al-Nar is designed to supply and subsidize products in the six emirates outside Abu Dhabi, which have been dependent on imports distributed by Shell, British Petroleum, and Caltex. These functions, however, are being taken over in 1981 by the new Emirates General Petroleum Corporation. Limited managerial and technical expertise, along with a small population, is likely to constrain the growth of Abu Dhabi National Oil Company (ADNOC) and other enterprises in the UAE in the future.

Iran

Iran's production depends on revenue needs, political decisions, maintenance and organization in the oil fields, and investment in surface facilities. In order to satisfy its current internal requirements and obtain sufficient foreign exchange earnings, Iran needs to lift no more than about 2.3 million b/d. This low level of production is possible because Iran's import requirements have dropped sharply and oil prices have more than doubled since 1978.

The Revolutionary Council sets production targets. In February 1979, Iran set 4 million b/d as the target for output, of which 0.7 million b/d were needed for internal refining and 3.25 million b/d were required for export. The target for 1980 was 3 to 3.5 million b/d, according to Oil Minister Moinfar, but President Bani-Sadr has indicated a preference for even lower output.

Another factor affecting Iranian output is technical: maintenance and organization in the oil fields. Labor unrest and recent bureaucratic and political upheavals have led many experienced technicians and managers to leave the country or retire, thus reducing Iran's ability to operate and manage its energy sector. The primary reservoirs cannot be damaged easily by current reduced production levels, but wells and surface facilities are maintenance-sensitive, and replacement would require capital investment, engineering resources, and long lead times to restore production. Iran can lift around 4 million b/d in 1980, but that capacity is declining at least 10 percent per year and possibly faster as a result of deteriorating surface facilities, not to mention physical plant destruction arising form the Iraq-Iran war that erupted in the autumn of 1980.

My assessment is that Iran will not produce more than 3 million b/d in 1980, and probably less. Furthermore, productive capacity will continue to decline until the Iranian regime becomes more stable. In the unlikely circumstance that the current political situation does not stabilize and improve, Iran's export capacity will be very small within five years. If and when the revolution moderates and NIOC regains more control over its operations, Iran can be expected to regain its important role in world oil trade.

Prior to the revolution, NIOC was the most developed national company in downstream operations. With a share in North Sea production, three refineries outside Iran, and direct sales that occasionally reached 1 million b/d, NIOC had developed substantial managerial and technical expertise. The political upheavals in 1979 and 1980 purged many of the most senior managers and technicians. NIOC will feel the effects of these personnel changes for the next decade. Still, the company has pioneered new contract terms that have influenced other nations.

Iraq

Iraq is OPEC's second largest oil producer and has extremely good potential for adding to its known reserves of 40 billion barrels. A conservative oil policy has been followed for many years, producing only slightly more oil than was necessary to meet foreign exchange requirements and build moderate reserves of foreign exchange. Iraq abandoned this policy in 1979, lifting far more crude than it needed to fund its imports and carry out a moderate development program. For 1980, Iraq has been very ambiguous about its intended production, with mixed signals emerging from Baghdad. The 1980-1981 political and military struggle with Iran may increase production as Iraq seeks to take over Iranian markets.

Iraq needs to produce only about 1.5 to 2 million b/d to satisfy all of its foreign exchange requirements; it raised oil revenues from $30 billion to $47 billion in 1979. The best estimate, however, is that Iraq will maintain output at 3.3-3.4 million b/d and that the country will continue to expand its oil-producing capacity with exploration and production arrangements with Western firms.

Sustainable capacity is currently about 3.8 million b/d and could rise to perhaps 4-4.3 million b/d by 1985 (with expansion at North Rumaila, Majnoon, and fields east of Baghdad offsetting declines at Kirkuk and possibly South Rumaila). This increase would confirm Iraq's position as the second largest producer in OPEC. The expanded capacity also would give Iraq increased advantage over Iran. For example, in return for various political concessions, Iraq might be willing to supply additional quantities of oil. The nation wants to enhance its position in the Arabian/Persian Gulf area, and increased oil-producing capacity and selective use of this capacity is one way to do it. One may conclude that Iraqi output will be about 3 to 4 million b/d in 1990. The actual will depend heavily on political factors, since any level of production in the just-noted range will produce more foreign exchange than Iraq is likely to need.

Iraq has not made any major moves in the direction of non-domestic downstream integration. Long a proponent of direct government-to-government sales, Iraq can be expected to increase sales in this area as well as tie crude to further exploration activity in Iraq. With a likely spare capacity of 1 million b/d or more, Iraq has some flexibility in negotiating favorable side arrangements for incremental crude.

Venezuela

Venezuelan oil output has declined steadily since 1973 to its current level of 2.3 million b/d and may well continue to fall through 1985. This decline is the result of low investment in exploration and development during the last 15 years. In 1979, a new 10-year plan was developed to substantially increase exploration and development expenditures. The $15 billion program is designed to increase capacity to 2.8 million b/d. Although the new exploration scheme appears promising, it will be the mid-1980s before the results are likely to appear.

Venezuela views the Orinoco Heavy Oil Belt as its long-term hedge for continued oil exports as reserves of light and medium crude oil decline (proven reserves as of January 1, 1981, were some 17.95 billion barrels). Petroven, the national oil company, has announced plans to produce and refine 125,000 b/d from Orinoco by 1985 and tentatively expects to add another 125,000 b/d by 1991. The long-term goal for heavy oil production is 1 million b/d by the end of the century.

Petroven is also deeply involved and interested in further downstream activity in marketing and refining. Resource constraints should hold production stable in the 1980s, and Venezuela is likely to maximize production at 2 to 2.2 million b/d through 1985. Their success will depend on the $15 billion 10-year development plan that is aimed at increasing Venezuelan capacity to 2.8 million b/d by 1988. Some $3 billion is designated for refinery modification that will increase gasoline yields and will adapt refineries for heavy crude throughput. With its large population, Venezuela is likely to maximize production.

Libya

Libya's oil production is likely to continue near its current level of 2 million b/d through 1985. After peaking at more than 3 million b/d in 1970, Libyan output has declined sharply under government-imposed quotas, poor company-government relations, and low investment levels in exploration and development. The potential for new discoveries does exist, and Libya is making some effort to attract foreign companies into exploration. However, without a major effort by Western companies, which probably requires better terms and improved political relations, it is questionable whether Libyan production can rise substantially.

The main determinants of Libyan oil policies are a mixture of geological/technical factors and political considerations. The country has used its oil income to support domestic economic development and foreign policy objectives. For domestic purposes (excluding arms), Libya probably needs to produce only around 1 million b/d. Lack of investment, however, is gradually reducing productive capacity. Given this declining capacity, Libya has attempted to maximize revenue by adopting a "hawk" position on oil prices and by lifting oil near capacity. It is unlikely that there will be a change in price policy, and production will probably continue at close to capacity except during any temporary political crisis. Libya announced a cutback from 2 million b/d in 1979 to 1.75 million b/d for the second quarter of 1980 in order to reinforce price advances by keeping the world oil market from experiencing a glut of significant proportions. In 1990 production may be 1.7 to 2.4 million b/d; the exact figure will depend largely on the success of any exploration.

It has been reported that Libya has drafted an exploration and production-sharing agreement that varies from an 85/15 percent to a 75/25 percent split. It is also said that 12 companies have reached tentative agreement for an overall billion-dollar exploration effort. Libya has been active in direct marketing and government-to-government crude sales, and with a prime location near European markets, Libya could easily expand refining and downstream activities to a larger scale.

Nigeria

Nigeria's proved oil reserves total 12.2 billion barrels, with an estimated 7 to 18 billion barrels of undiscovered oil in place. The oil is found in small to medium-sized fields, and the Nigerian National Oil Corporation (NNOC) is pushing very hard for an expanded exploration and development effort with Western companies. Revenues increased considerably in 1979 but Nigeria's population (80 to 100 million) dictates that Nigeria maximize both price and production (which is expected to modestly increase to 2.5 million b/d by 1985). Nigeria has not made any major move downstream. NNOC has placed a cap on contract sales, however, and is attempting to market oil in the spot market and above the official sales price. One constraint is that Nigerian officials remember events of early 1977 when they were caught in a mini-glut with their prices high. Nigeria has comparatively few options, and no major changes are expected in marketing or production plans.

Algeria

Algeria is long on experience and technical capabilities but short on oil reserves, which total 7 billion barrels. The country has the capacity to produce 1.2 million b/d, which could be held even with near-term exploration successes. It is reported that crude contract customers have been given until October 1, 1980, to submit exploration and production proposals or risk losing oil. The agreements call for training and research as well as equity participation of between 30 and 49 percent with Sonatrach (the national oil company). Firms that meet Sonatrach's requirements will receive a $3 per barrel reduction in the cost of crude. Algeria has significantly increased refining and gas-processing capabilities and is about to capitalize on these important aspects. It can be expected that Algeria will further the development of its operations in downstream activities.

Mexico

From the U.S. perspective, Mexico, by virtue of its physical proximity and close economic ties to the United States as well as its large resource base, represents a particularly attractive source of strategically secure oil and gas. Therefore, the United States clearly aims to encourage the expansion of Mexican oil and gas production. However, it should be recognized that decisions regarding the timing and levels of Mexican energy development will be made consistent with that government's energy and economic goals and with an acute sensitivity to U.S. influence in the development of the Mexican patrimony.

Current production averages about 1.9 million barrels of crude oil per day, over half of which is produced in the Reforma fields. Approximately 700,000 b/d are exported, with 80 percent shipped to U.S. markets. Future production levels will be based on political as well as socioeconomic considerations. Paramount among these considerations are the impact of increased output on income distribution, the demands of a growing population, and the existence of a current account deficit. All of these factors support increased exports. However, there is also considerable feeling in Mexico for conserving the country's national patrimony: oil and gas reserves.

PEMEX, the national oil company, is responsible for developing Mexico's petroleum reserves. The original PEMEX Six-Year Plan (1976-1982) called for production to increase to 2.25 million b/d by the end of 1980, with projected 1982 production levels in excess of 3 million b/d. The 2.25 million b/d target will probably be met by June of 1980, and consequent 1982 production levels could reach 3.5 million b/d, with 1.4 to 2.1 million b/d available for export. By 1985, the doubling of production to 4 million b/d would be possible, with about half of it destined for export. Nonetheless, there have been some reports that Mexico has been rethinking production targets. PEMEX can be expected to expand its downstream role in refining, transportation, and marketing in the Western hemisphere. With some 40 years of experience, PEMEX has the technical know-how to increase activities in this area.

People's Republic of China

Over the course of the past few years, the People's Republic of China

(PRC) has emerged as the most important new exploration area in the world. Over half of the PRC's 30 hydrocarbon basins remain relatively unexplored. Offshore basins are particularly attractive, and extensive geophysical work is currently under way. As exploratory drilling is expanded, the PRC could prove to hold one of the larger oil reserves in the world. Proved and probable reserves approximate 15 billion barrels, with potential for up to 72 to 100 billion barrels.

The PRC currently produces approximately 2.1 million b/d, and output is expected to grow conservatively at 2.5 percent per year through 1987. Increases beyond that time will depend on the rate and success of future oil exploration ventures, which are scheduled to begin in 1981. Still, with a large number of attractive exploration areas, particularly offshore, China could significantly expand production by the end of the decade if early exploration efforts go well. Increased domestic demand will be a major constraint on exportable volumes, however, with exports expected to remain modest (they are currently 332,000 b/d of both crude and product).

Conclusions

Some general observations can be drawn: (a) in production, host producer governments are likely to maximize the use of oil revenues for domestic social, economic, and political purposes rather than reinvesting in the industry; (b) this will further accelerate tensions between producer governments and their national oil companies; (c) investment in the operational aspects of the various industries will suffer, causing further production constraints; (d) at the same time many producer countries will maximize oil production and prices.

Turning to the marketing element in the equation: (a) some of the more advanced national companies will further expand into downstream operations, others will be constrained; (b) the movement, however, will be in the direction of greater downstream operations – some will succeed and some will fail; (c) expansion will take place even though the economics are not good. The investment of excess revenues, expanding indigenous technical work forces, and politics and national pride will often overrule the economic arguments and skepticism.

The modern oil industry developed its structure in the 1920s and was shaped by visionaries and clairvoyants who went against conventional wisdom. The 1980s will very likely be a comparable period.

10

DOWNSTREAM OPERATIONS AND THE DEVELOPMENT OF OPEC MEMBER COUNTRIES

*Ali M. Jaidah**

Time, it is said, is the enemy of us all. In the case of OPEC, these are no empty words, but indicative of our true situation. We in OPEC are indeed engaged in a race against time to develop our economies, and it is for this reason — and no other — that the member countries have turned their attention to downstream operations.

Of course, there are those who would criticize OPEC states for the desire to move into a field which is already, so to speak, "overpopulated," in other words, suffering from considerable overcapacity on a world scale. These critics are to be found both outside and within OPEC. The former may fear competition with them; the latter — some of the member countries — are worried about the social and economic implications involved in such a move and feel that the major oil exporters should seek some other avenue of industrial expansion.

Thus, it was the objective of the October 1978 seminar sponsored by OPEC to offer a platform for the airing of all points of view on this most important issue; OPEC states have reached a decisive crossroads and must plan now for the path ahead. In one respect, however, OPEC member countries are unanimous: they must break free from their dependence on a single marketable commodity; they can no longer afford to play the role of mere residual suppliers of a raw material.

---

*Ali Mohammed Jaidah earned undergraduate and graduate degrees in economics and petroleum economics from the University of London; prior to his present position as Director General of the Qatar General Petroleum Corporation (QGPC), he served as Secretary General of the Organization of the Petroleum Exporting Countries (January 1977 to December 1978). The author earlier was: Director of Petroleum Affairs, Ministry of Finance and Petroleum of Qatar; Member of the Board of Directors, QGPC; Qatar Governor for OPEC; Member of the Executive Office of OAPEC (Organization of the Arab Petroleum Exporting Countries); as well as head of Qatari delegations to OPEC, OAPEC, and various other conferences and meetings.

The Development Impetus

OPEC nations have basic economic characteristics common to all developing countries. They are familiar to many readers. OPEC states do not possess the technological know-how and the necessary pre-requisites for economic and industrial take-off; neither do these countries have a comprehensive structure for sustainable growth. Their present well-being — and this is what is disturbing when looking into the future — is derived from the export of a single raw material, finite in its nature, a resource which will be exhausted within the next three or four decades.

What then? Are the countries in OPEC to repeat the historical cycle epitomized by Spain's "Golden Age" of sudden but not lasting wealth fueled from the New World? It would be tragic if this were to happen. This brings us back to the point made earlier: time, for the OPEC nations, is crucial. Their policy makers and planners seek — and must find — means of attaining a reasonable standard of living for future generations; they must act now to provide against the day when there is no more oil and gas. In order to do this, they have to create viable economic infrastructures on which the member countries can build a sound and secure economic base.

As well known, the early industrialization of Europe and North America was based on the existence of indigenous coal and iron ore deposits. The main industrial centers which gave momentum to the Industrial Revolution of the late eighteenth and early nineteenth centuries were founded around deposits of such natural resources. Subsequent industrial development can also be traced, in many cases, to the exploitation of indigenous natural resources where there existed the necessary infrastructure.

Most of the present-day industrial nations import raw materials from less-developed areas. It is no accident that underdevelopment is synonymous with a state of affairs characterized by dependence on raw material exports, while industrial advancement is typified by the full utilization of indigenous resources, plus a measure of raw material imports. In the light of this pattern of development, is it not natural that OPEC member countries, in striving to catch up with the industrialized nations should adopt the quickest and most immediately available course open to them, by following the early example of the present industrial nations in seeking to process their own indigenous hydrocarbon deposits in whatever ways possible? Indeed, the exporting countries have no other economic advantage; the raw material is to be found on their doorstep, offering the logical starting point.

Of course, it is true that as a result of massive exports of crude oil, OPEC states have earned in recent years relatively large amounts of foreign exchange. It is also the case that this has given them an advantage which most of the other developing countries lack. However, it must never be forgotten that the wealth presently accruing to OPEC member countries in the shape of foreign exchange revenues is of a transitory nature; OPEC nations are not, in any lasting sense, more developed that the rest of the Third World.

This is why they must seize the opportunity presented by the oil revenues to transform significant parts of this foreign exchange income into capital formation of a nature which will lead them into a state of self-sustained economic development and growth. However, since many OPEC countries have no apparent national endowments other than the availability of liquid and gas hydrocarbon deposits, their options for substantial investment outside the field of processing these deposits are few, with the exception, of course, of investment in the development of human resources — an investment which, in this author's view, should have top priority among the existing possibilities since human resources are most critical factors in development.

Hopefully it has been made quite clear that the overriding national interest of most OPEC member countries leaves their planners little choice but to take the risk of moving downstream in refining, petrochemicals and other related activities. Building refineries and fertilizers plants for domestic markets constitutes a relatively minor contribution to the degree of investment necessary to bring about a speedy take-off of the economy as a whole.

Unfortunately, the experience of OPEC member countries since the Second World War has shown that the industrialized consuming countries are not prepared to encourage raw material producers to process their own raw materials. Quite deliberately, refineries were built, not beside the resource, but in locations inside Europe, Japan, and North America. One can only deduce from all this that the present resistance of OECD nations to downstream ventures on the part of OPEC states stems, not so much from the existence of surplus capacities in such industries, but rather from a fear of competition in the long run.

The Rationale for Moving Downstream

It is often argued that the main parameters on which industrial investment in developing countries is based can be found in one or more of the following: (1) to utilize an indigenous raw material and to recuperate the value added; (2) to create new employment opportunities; (3) to help in the learning process and the dissemination of new technologies; (4) to substitute for imports from abroad; (5) to encourage development of new regions of the country concerned and to induce further related industrial activities; and (6) to reduce the heavy dependence on foreign supplies and to create a measure of self-reliance.

It is significant that none of the above considerations of national interest is based purely on simple feasibility criteria. It should not be regarded as surprising, therefore, if the same attitude is taken by OPEC countries with regard to downstream operations. On the other hand, one very important factor here must always be kept in mind: even if for some OPEC states projects appear economically feasible on paper, the implementation of these projects could impose a heavy burden on the economic and social structure of their indigenous societies. It must be frankly stated that there are cases — countries like Kuwait, the United Arab Emirates, and Qatar — where, in the long run, the need on the part of the country to import labor to operate plants and to provide that labor with free education, health and heavily subsidized electricity and water facilities can outweigh the advantages of the projects themselves. In addition, regard must be paid to the social consequences of introducing a foreign labor force to a small country, a labor force which, in some cases, might exceed the native population. Those conversant with the OPEC nations are familiar with the tensions and political problems which can arise in such a situation.

Despite the risks involved, however, all of the major oil-exporting countries are anxious to go downstream for the reasons already mentioned. This writer has always been a strong advocate of their legitimate right to do so. However, each member country must be analyzed separately in the light of its own peculiar capacity and requirements. There is no doubt, of course, that refining for local consumption, as well as some extra capacity for exports, cannot be argued against. It is not logical, for example, to expect Nigeria and Indonesia to continue to import petroleum products simply because surplus capacities exist abroad. It is also natural that agricultural countries, such as Iran, should build fertilizer plants regardless of capacities abroad if they can use most of the production at home.

Still, one may conclude that no general case can be made for *all* OPEC countries to go downstream at *all* costs. In the case of some members, it may be found advisable to opt for less-complicated technology in the initial stage with careful planning and development of human resources in order to involve the indigenous population at all stages from construction to operation. This would allow moving away from the present system whereby, in some member countries, a foreign company builds the plant on a turn-key basis, and the nation in turn brings foreign labor to man it, pays all capital requirements, provides the infrastructure free of charge, supplies the raw materials on preferential terms, and pays management fees; on top of all this, the country has to provide rescue if the venture goes wrong. The benefits to the rest of the economy can be negligible, since, for example, wages are transferred abroad. All that is derived is a probability of some return on investment — something which could be earned simply through bank deposits.

This situation is particularly true of joint ventures where the oil-exporting countries bear all the financial burdens and most of the risks while the foreign partner gets guaranteed marketing and management fees and ties down the plant in long-term arrangements for operation, maintenance, spare parts and even, in some cases, royalties. Where, in all this, is there room for the application of the learning process and a meaningful transfer of technology? If OPEC nations are to forge ahead industrially, this state of affairs must be altered.

OPEC Downstream Operations Today: An Assessment

Meanwhile, the oil-exporting nations have not been standing still. Since the coming to maturity of most of their national oil companies (NOCs) much has been achieved, despite limited experience, in the progressive take-over of the production stage of oil and natural gas. With increasing confidence, the NOCs are venturing into new avenues of development. Since they are now directly involved in the production of crude oil, there is a growing awareness on their part of the need to put an end to the utter waste involved in the flaring of associated gas. This has led to rapid growth of investment in efforts to achieve the complete utilization of associated gas—in reinjection, in exports, in processing and petrochemicals and, significantly, through intensified domestic consumption as a substitute for other fuels which might be more remunerative as exports.

The members of OPEC have never concealed their readiness to learn from the former controlling power in the oil industry (the majors), and, not surprisingly, the countries are finding that the path taken long ago by the companies in adopting an integrated approach to the industry is also a logical path to follow. Some NOCs are venturing into refining and products' marketing alongside their crude oil export operations. Several NOCs have their own tanker fleets. Surely it is only to be expected that the mere existence of national oil companies will, in itself, lead to ventures in downstream operations and related activities, following the example already set by the multinationals. This involvement is intensified by the fact that OPEC NOCs naturally feel that it is their duty to assist their countries' development efforts.

Unfortunately, the industrialized states do not appear to share OPEC aspirations in this regard. Of course, this is nothing new. Throughout its 18-year history, OPEC's *raison d'etre* has been disputed, its objectives often deliberately misconstrued. Indeed, this author is by no means sure that the exporter organization has yet succeeded in establishing in the industrialized world an "acceptable face of OPEC." However that may be, there is a commitment to pursue the chosen course in the belief that it is a moral duty to do so for the benefit of future generations in OPEC countries.

Intelligence is by no means a prerogative of industrialized societies, and what does especially upset OPEC members is being told they are wrong before they have even had the chance to say what they want. If OPEC states earn money, then their assets, it is said, are destroying the monetary system. If the same countries go downstream, then it is said they are damaging everybody's interests.

This seems very one-sided. Seldom if ever does one hear a sympathetic word about *our* interests, *our* needs, *our* legitimate aspirations. Everything the member countries have achieved thus far has had to be won without help, indeed, usually against the most determined opposition. Now there is criticism for seeking to penetrate the downstream sector of the industry, despite the fact that the reasons for doing so are clear and the expressions by OPEC states have been for goodwill and cooperation. The October 1978 OPEC seminar and discussions raised by articles and papers hopefully will contribute to a fundamental change in that attitude.

Having dealt in general terms with the motivation behind the member countries' desire to go downstream, it might be helpful to underline OPEC's case with a few comparisons. OPEC nations comprise 320 million people, or nearly 8 percent of the world's population. The oil states grouped within OPEC command 68 percent of estimated total world reserves of crude oil; their share of total world crude oil production stands at 52 percent; and their crude oil exports account for 84 percent of the world total. OPEC members hold 39 percent of total world reserves of natural gas and supply 16 percent of worldwide natural gas production.

In contrast to the above, OPEC's share of total world refining capacity (which in 1977 reached 76.7 million barrels per day) is a meager 6 percent (although, in 1977, member countries' refinery capacity, on a volumetric basis, increased by 4.6 percent over the previous year). OPEC's share of the world petrochemical industry is only 3.2 percent and a mere 2.9 percent of its crude exports are transported in OPEC tankers. Such statistics show how very far OPEC states have still to go before they achieve anything like the market share to which their contribution to the oil industry and the world economy entitles them. Until a more equal balance is reached in the downstream sector, the member countries can never hope to realize their aspirations in development and industrialization.

The figures are indeed disturbing. They indicate that OPEC refining capacity has only increased by a mere 600,000 barrels per day, or 14 percent between 1973 and 1977, which is less than the increase in OPEC domestic demand during the same period. Additionally, it is also obvious that OPEC product exports were equal to only about 7 percent of its total crude exports.

This is in contrast to the fact that the industrial areas have increased their share of world refining capacity from 77.7 percent to 80.4 percent in about the same period.

In the petrochemical field, OPEC's small share of 3.2 percent was due to the fact that in most member countries the petrochemical industry either did not exist at all or was, at best, negligible, the exceptions being Venezuela, Iran, and Kuwait which together produce more than two-thirds of the total OPEC petrochemical output. Fairly large petrochemical complexes are however presently in the design stage or under construction in Qatar, Algeria, Ecuador, Libya, Iran, Iraq, and Saudi Arabia.

Downstream Options for OPEC

It must be noted that vast divergences can be observed in the various member countries with regard to their economic, political and social structures. They vary according to their history and experience in the oil industry, the size of their population, their agricultural base, the extent of their industrial development and availability of infrastructure, their oil and gas reserves and ultimate hydrocarbon potential, their geographical position vis-a-vis the major consuming markets, their balance-of-payments position and command of capital. All of these factors modify and constrain the opportunity for development and make it difficult to offer a comment of general applicability. Equally, their aspirations to go downstream and the type and degree of involvement they envisage vary, depending upon the structure and condition of the country. However, one can place the OPEC states in two broad categories, namely: (a) countries with an appreciable domestic market that could actually consume a high percentage of the products produced, hence giving an economic and social justification for going downstream, and (b) nations with no appreciable domestic market, hence their downstream operations would be mainly export oriented and, even within this category, the geographical location with respect to export markets plays an import role in the profitability of their operation.

What then can be said of the much-publicized impediments to OPEC's going downstream? These can be summarized as follow. First, due to the nonavailability of domestic markets, OPEC's downstream operations have to be export oriented and in a situation of excess surplus capacity and a position of nonprofitability. This would bring further disequilibrium and chaos to the market. Additionally, these export oriented products have either to cater to a special demand pattern in a particular country or region and, thereby, be totally dependent upon that market, or alternatively, produce a whole range of products with different specifications that would cater to multiple markets, thereby increasing the difficulties and raising the production costs. Second, there is the high cost of importing technology and materials. The lack of expertise to supervise, operate and market these products stands as the third impediment. Additional factors are the high cost of transporation, the nonavailability of the required infrastructure, and the distortion of the petroleum market and adherence to the price structure (as the marketing of products is more competitive than the market of crudes), thereby increasing the possibility of price undercutting. Seventh, these highly technical projects are less profitable, very capital intensive, and low labor intensive. Finally, the dependence on world markets will disrupt economic integration within the country or its regional integration. Given such conditions, what are the options available to OPEC? In trying to answer this question, let us view some of the factors influencing OPEC's involvement in downstream operations.

The often publicized and frequent comment concerning the underutilization of refinery capacity generally only refers to primary distillation and does not include secondary processes which currently are fully utilized worldwide, resulting in an erroneous picture. Additionally, it is worth mentioning that the profit of the refiners is governed primarily by the severity of cracking operations, not the loading of crude unit. The prevailing surplus capacity in distillation is expected to disappear in the near future as the demand for refined products in the major consuming countries is projected to increase with the biggest share being taken by the light ends and, especially, petrochemical naphtha. Forecasts envisage such increase up to the year 1990, with the exception of the demand for motor gasoline in the United States. Moreover, the expected changes in the chemistry of the crude supply and the chemistry of the demand will necessitate either modification of existing refinery configurations or expansion, especially in the major secondary processing units. This is due to the fact that tighter fuel specifications to comply with environmental protection measures will be set, and alternative sources of energy will displace some of the lower end of the barrel. It is logical, therefore, that OPEC nations should be allowed to take their rightful share of the refining industry and benefit from this trend in demand through the construction of refining facilities in their own countries.

Regarding the petrochemical issue and bearing in mind that additional capacity is being built or planned even though the 1976 capacity is sufficient to satisfy the projected 1981 demand, it could be reasoned that this apparent contradiction could be due to any of the following factors: (a) the nature of the petrochemical industry necessitates building of excess capacity for greater flexibility of utilization; (b) the forecast demand indicated is below that anticipated by those that have made, or are planning to make, an investment; and (c) to discourage others from building petrochemical complexes.

It should be clear from the context of this article that OPEC member countries have no alternative but to increase their involvement in downstream operations in order to pull themselves out of their state of underdevelopment within the lifetime of their hydrocarbon resources. The *modus operandi* and the forms which this involvement shall take must rest with the individual nations, each acting in accordance with its own particular circumstances and requirements.

We believe, however, that the most rational way to tackle the problem lies in a global approach based on a reasonable appreciation by the developed countries of the needs, requirements, and aspirations of the producers and by the latter, of the essential needs of the consuming countries. It has never been OPEC's intention to disrupt the international oil industry; it is inconceivable that the organization should want to damage the economies of its traditional trading partners. OPEC's policies since assuming control of prices and production have been notable for their moderation and consideration for the economic problems of the major consuming countries.

At the same time, however, it should be emphasized that OPEC feels the developed countries have a reciprocal obligation to assist the development process in the member nations through the adequate and timely transfer of modern technology and by opening up market opportunities for OPEC products. The industrialized world must realize that the oil exporters cannot continue indefinitely to pay the exorbitant cost of imported equipment and know-how. Furthermore, OPEC members must be allowed reasonable access to the market for finished and semi-finished products. It is not equitable that the major petroleum producers should buy equipment and spare parts from the industrialized countries, as well as importing their skilled labor, only to be deprived of the right to sell those products again in the advanced-nation markets. This state of affairs, if it were to continue, might compel the member countries to tie their crude sales to the sales of finished products.

In conclusion, we should recognize that OPEC member countries have already started out along the road towards total involvement in downstream operations; they cannot turn back now. Hopefully, OPEC states will not be forced to act unilaterally in this sphere. As in the past, the oil exporters seek an accommodation with the developed countries which will permit a smooth and harmonious transition from the role of mere residual suppliers of raw materials to that of full and equal partners in all activities of the international oil industry.

11

OPEC AID, THE OPEC FUND, AND COOPERATION WITH COMMERCIAL DEVELOPMENT FINANCE SOURCES

*Ibrahim F. I. Shihata**

Among the concessionary sources of external finance available to developing countries, the OPEC Special Fund is probably the newest on the scene. It is, however, the one source that reflects the joint aid efforts of *all* OPEC members, that is, of practically all donor developing nations. Equally important, it is one of the sources keenly interested in the search for new solutions to the development problem. Not only does it work under more liberal articles of agreement than those governing other official aid agencies, it is also free, or, if one may say so, innocent, of the "long-standing policies" which often inhibit older institutions from trying untested ideas. Since the OPEC Special Fund is the collective aid facility of OPEC members, it is presented here in a wider perspective. Description of its activities will be preceded by general remarks on the assistance rendered by OPEC members as a whole to other developing countries. The analysis of the Fund's role will also be followed by an elaboration of the role Western banks can and, in this author's view, *should* play in project financing within developing nations, especially in association with the concessionary sources of finance, the traditional as well as the new.

OPEC Aid

According to recently compiled United Nations figures, after a fourfold

---

*Ibrahim F. I. Shihata, presently Director-General of the OPEC Special Fund, holds degrees from Cairo University (LL.B, LL.M.) and Harvard University (S.J.D.). Prior to assuming his responsibilities at the Fund, he was a member of: the Conseil d'État in Cairo; the Technical Bureau of the President of the United Arab Republic in Damascus (1959-60); the faculty of the School of Law in Cairo University; the Kuwait Fund for Arab Economic Development as senior legal advisor and general counsel. Dr. Shihata is the author of 11 books and more than 50 articles published in English and Arabic in Europe, the Middle East, and the United States dealing with various aspects of international law and international development finance.

increase between 1973 and 1974, total *commitments* of OPEC donors to other developing countries increased by a further 22 percent in 1975 to reach a figure in excess of $15 billion, which corresponded then to 7.5 percent of the combined GNP (gross national product) of OPEC aid extenders.[1] The somewhat lower figures for 1976 and 1977 have remained, nonetheless, above the $10 billion mark. Actual net *disbursements* achieved, according to the same source, a near fivefold jump between 1973 and 1974 and were followed by a further 50 percent increase in 1975 for a total net outflow of $11.5 billion, corresponding to 5.6 percent of the donor's GNP (or to above 12 percent of GNP, if we restrict the data to the four major *Arab* extenders). As a result of these developments, OPEC states occupied the first six positions in ranking among all donor countries of the world as regards the proportion of aid to GNP in 1976.[2] Two of them ranked among the six largest bilateral donors in absolute terms, Saudi Arabia being second only to the United States.

Although there is no justification for comparing the aid performance of the industrialized, rich countries with that of a group of developing nations, a few of which may be in possession now of large net liquid assets but none of which is developed in the economic sense, such a comparison yields startling results. The figure for total commitments from the DAC (the Development Assistance Committee of the Organization for Economic Cooperation and Development or OECD composed of the United States, Canada, Japan, and Western European states) has been estimated around $25.3 billion in 1975. All official flows from these countries reached $16.5 billion in the same year and remained roughly unchanged in real terms in subsequent years, corresponding to less than 0.7 percent of the combined GNP of the donor countries.[3] Expressing this relationship differently, the commitments in 1975 of what amounts to less than a dozen OPEC donors represented more than 60 percent of the DAC total, although the combined GNP of countries in the latter group was almost 16 times as great as that of the former bloc. In 1977, OPEC members gave at least two and one-half times as much as the United States did. If DAC countries were

---

[1] United Nations, Conference on Trade and Development, "Financial Solidarity for Development: Efforts and Institutions of the Members of OPEC - 1977 Review" (draft of October 31, 1977, mimeographed, scheduled for publication).

[2] Disbursement of concessional flows (on ODA terms) from OPEC members to other developing countries has reached $1.1 billion in 1973, $3.5 billion in 1974, $5.2 billion in 1975, $5.2 billion in 1976, and $5.5 billion in 1977, corresponding roughly to a percentage of the GNP of the donors ranging from 1.5 to 2.7 percent.

[3] Net official disbursements (on ODA terms) from DAC countries reached in the same year $13.5 billion, rose to slightly above $13.6 billion in 1976, and reached $14.8 billion in 1977, corresponding to about 0.3 percent of their GNP, according to OECD sources.

to give concessional assistance in the same ratio of aid to GNP as do OPEC members, they would have extended more than tenfold what was actually expended in that year.

Against this gross disparity between the aid ratios of OPEC members and DAC countries (not to mention the much poorer performance of the Soviet bloc) are two noteworthy features which provide yet a more vivid contrast. First, OPEC assistance, unlike much of the aid provided by DAC nations, is untied to source while DAC flows are, almost inevitably, recycled back to the donors. One can say inevitably because even the untied DAC flows readily find their way back either to the donor country itself or to another Western economy, since DAC states constitute collectively by far the major source of procurement for the goods and services that recipients purchase. This, of course, results eventually in a much lower net aid transfer from the DAC extenders. On the other hand, OPEC aid brings no financial return to its donors. If anything, it actually helps two sources simultaneously: the developing, recipient countries as well as the developed nations from which procurements are made. (As a more cynical observer might put it, admittedly not without some exaggeration, aid may have its *recipients* in the developing world, but its *beneficiaries* are almost always in the developed world.) This distinguishing feature, coupled with the hard fact that OPEC aid is extended from revenues generated from exploitation of a commodity itself depletable (not from a renewable income yielded by industrial or agricultural production), makes the aid ratio comparisons even more disproportionate.

On the other hand, OPEC inflows, which are not and have not been meant to provide compensation for higher oil prices, accounted for the financing in 1975 of the equivalent of 99 percent of the value of net oil imports of other developing countries as a whole. This aid exceeded the increase of the aggregate oil bills of Sub-Saharan African nations while reaching two to three times the value of incremental oil imports of many of the least developed countries. In 1976, OPEC inflows financed directly up to 26 percent of the current account deficits of nonoil, developing states. In a broader sense, as the recent UNCTAD report utilized in this article confirms, "the entire deficit may be thought of as having been financed by OPEC capital exports recycled via the Eurocurrency and other financial markets."[4]

This OPEC aid is, of course, a relatively novel phenomenon in that it represents a transfer of resources from one group of *developing* countries to another. It is novel also because it has been swiftly extended through a broad range of channels combining bilateral, regional and multilateral

---

[4]United Nations, Conference on Trade and Development, op. cit.

sources. The OPEC Special Fund is just one among these channels. Although it has its own characteristics, its creation represents only the most recent of several initiatives taken by OPEC members to consolidate their position within and solidarity with the rest of the developing world.

The OPEC Special Fund

Constituted in 1976, the OPEC Special Fund is the aid facility of the 13 members of OPEC which have all signed its establishing agreement and contributed to its resources. The Fund began its operations in August 1976 with initial contributions of approximately $800 million. Subsequent contributions, some of which are earmarked for transfers to various international aid agencies, have brought the total contributions to date to something in excess of $1.6 billion. The transfer of funds to other agencies amounted to $481 million, thus making the net resources available to the OPEC Special Fund for its direct operations over $1.2 billion.

Basically the Fund is entrusted with two functions: that of coordinator of the joint OPEC members policies and activities in the field of external assistance and that of a collective aid donor in its own right.

Briefly, in its *role of coordinator* the Fund, or rather its Ministerial Committee and its Governing Committee, constitute the voice through which OPEC states have chosen to speak on various aspects of their external aid policies. Generally, the Fund has tried since its foundation to play a positive role in supporting new institutions and in developing appropriate policies to effect changes in international economic relations. The main objective is to achieve the establishment of the "New International Economic Order" through a maximum of action and a minimum of rhetoric. The participation of the Fund in the creation of the International Fund for Agricultural Development (IFAD), for example, was an effort in partial fulfillment of this objective. OPEC members contributed through the Fund $435.5 million (or almost half the initial capital subscriptions to that newly created institution), thus securing two-thirds of the voting power in IFAD's governing bodies for developing countries. Another instance is the interest shown by OPEC states in UNCTAD's Integrated Program for Commodities and its main feature, the projected Common Fund. The primary purpose of the Common Fund is to provide assistance to individual commodity organizations for purchase and stocking of commodities when prices fall below an agreed upon floor. Price support would then be provided when supply and demand conditions adversely threaten price and other economic effects to producers. The OPEC Special Fund was entrusted with the responsibility of coordinating the efforts of OPEC members in the negotiations for the establishment of the Common Fund. These countries' financial contribution to the Common Fund will also be made eventually through the OPEC Special Fund. In another development, the Ministerial Committee, which oversees the activities of the Fund, was instrumental in arranging for the donation of the profits accruing to a number of OPEC states from the International Monetary Fund (IMF) gold sales to the OPEC Special Fund, which in turn donated these profits to the Trust Fund administered by the IMF. An arrangement with the United Nations Development Fund (UNDP), also in the form of a collective donation through the Fund, has been made to finance a number of the UNDP technical assistance regional projects. Yet another important facet of the Fund's role as coordinator concerns the activities of *national* aid agencies of the individual OPEC states. The task of coordinating the policies and programs of these agencies is now entrusted to the Governing Committee of the OPEC Special Fund which has just started to play an active role in this field as well.

The Fund's major *role as aid donor* on highly concessional terms is, of course, the main reason for its coming into being. The Fund conducts basically two sets of lending activities: (1) it provides balance-of-payments assistance to countries with severe deficits in their current accounts and (2) it extends loans for economic development projects. All Fund loans have long maturities and all but a few of them have been granted free of interest.

Balance-of-payments (BOP) loans, 60 of which have so far been made, are now generally extended on a smaller scale than in the past. At present, two criteria must be satisfied by an applicant country to benefit from this type of assistance: (a) the current deficit factor just mentioned and (b) the absence of quick-disbursing projects suitable for financing. Generally, when the Fund extends a BOP loan, the borrower (which in all cases is a government) agrees to deposit an equivalent amount in local currency to be used within a reasonable period of time for local cost financing of one or more development projects. When no such use is made of counterpart funds mobilized under BOP loans, the maturity of the loan itself is considerably reduced. This shortening of the loan repayment period is, in effect, an incentive to the governments concerned to mobilize domestic financial resources for the internal financing of projects that may not otherwise be realized. By and large, the Fund's experience with this type of assistance has been successful thus far as loan proceeds have been swiftly disbursed and, in the vast majority of cases, governments have opted to mobilize and use local counterpart funds for domestic projects. In fact, to date some 65 projects in 37 countries have been approved for financing with local funds totaling $160 million. Most of these projects were cofinanced by other external sources and a few of them have benefited from additional *foreign exchange* financing from the Fund itself.

The second type of assistance extended by the Fund is in the form of direct project lending which has now become the Fund's major operational activity. This type of undertaking is carried out under a philosophy expressly propounded by the authors of the agreement establishing the Fund. They have taken great care to avoid creating another bureaucracy, duplicating organizational structures and activities of other institutions. Rather, they have sought to create an institution empowered with swift decision-making capabilities, while maintaining a lean staffing structure to avoid the bottlenecks and delays so characteristic of larger, more formally structured bodies. Typically, when a well-studied project is submitted for the Fund's consideration, the time lag between receipt of the documentation and final approval is rarely more than three months. Some of the projects the Fund cofinances may be nonrevenue earning and essentially social in character, or they may be of a commercial nature. In all cases, economic and social criteria are the paramount considerations in the decision to finance a project. Technical considerations, of course, are taken into account to evaluate the soundness of a project and in the case of revenue-earning projects, particularly those of a commercial nature, assured financial viability is also a requisite for the go-ahead decision. To date, i.e., the close of 1978, the Fund has committed $378 million for the financing of 107 projects (including the 64 projects financed by local counterpart funds). These commitments under our project loan portfolio have covered some 61 countries in the developing world, with even "repeater" projects in some of them, despite the relatively short period in which the Fund has been in operation. The projects financed have covered practically all economic sectors with power having the largest share (30 percent) followed by industry (27.3 percent), agriculture (18 percent), transport (14 percent), public utilities (9.5 percent), and telecommunication (1.2 percent). Some of these projects are self-liquidating with their debt obligations serviced from cash-flow earnings; others are so-called "green-field" investments, that is, of a commercial nature but requiring complementary financing for basic infrastructure support, but most are simply infrastructure projects.

Project-related assistance has been extended swiftly, as attested by the number of projects so far financed. This is largely a result of the close *working relationship* established by the Fund with almost all the international development aid agencies, be they worldwide or of regional character, as well as with national aid agencies belonging to OPEC members. The essence of this relationship was laid down in the agreement establishing the Fund which specified that appraisal of projects to be financed by the Fund must be undertaken by "an appropriate international agency or by an agency of a member country." Again, the authors of the agreement, being fully aware of the danger of duplicating efforts in the field of international development assistance, directed the Fund to draw heavily on the talent and services of other established agencies for the administration of its loans. Implicitly, this has meant that the Fund would give priority to cofinancing arrangements and, in fact, this has been the case in practice.[5] The advantages of cofinancing arrangements are not restricted to the Fund's considerable savings in time and cost, although it should be mentioned here that the Fund incurs no cost for the appraisal and project-monitoring activities carried out by other institutions on its behalf. More importantly, cofinancing allows for the immediate implementation of projects which might otherwise be indefinitely postponed until the financing gap could be filled. As is well known, such a postponement, if not avoided, almost always entails a widening of the very gap it was meant to help bridge due to cost overruns. For the Fund, this role of "gap financier" also makes it, in fact, a lender-of-last-resort and thus unlikely to be found in competition with other sources of finance for a given project. Being drawn in rather late in the project preparation cycle, the Fund is in a better position to give objective consideration to the merits of projects submitted for its partial financing. Moreover, the Fund has also been involved occasionally in the partial financing of "virgin" projects where it played a catalytic role seeking complementary financing from other sources, especially those to which it is closely linked.

Once a go-ahead decision for a particular project is made, cooperation with the other aid agency which has prepared and appraised the project is not terminated but extends to cover the project implementation period. The framework of this cooperation at the postinvestment stage is detailed in a letter of cooperation signed by the Fund and countersigned by the other aid agency which, in so doing, assumes the responsibility for administering the loan on behalf of the Fund. The first such cooperation was implemented in conjunction with the World Bank, and it now exists with almost all regional aid agencies and with practically all the national development funds of OPEC members. In each case, the loan administrator conducts the normal supervisory work required to monitor the overall progress of project implementation with the same diligence it would in administering its own loans. The Fund maintains the right to participate in supervisory missions for which it receives, in any event, both the terms of reference and the project progress reports. The loan administrator reviews withdrawal

---

[5]Cofinancing is now a general phenomenon in OPEC assistance. According to OECD sources, by the end of March 1978, up to 106 projects (including projects in an advanced stage of negotiations) have been financed jointly by OPEC and non-OPEC sources. The total amount involved was roughly $7.3 billion of which some $3 billion were from OPEC sources. Ibrahim Shihata, "The Working Relationship between the OPEC Special Fund and Other Development Financing Institutions," *OPEC Review*, June 1978, pp. 8-14, offers details on supported projects and procedures.

applications for the Fund loans and issues appropriate consent on which basis the Fund directly handles the disbursement process. Usually, there is provision for consultation on a regular basis among cofinanciers and both parties agree to refrain from such unusual action as suspension or cancellation of the loan without mutual consultation. From the borrower's viewpoint, the absence of effort duplication under such loan administration arrangements translates into obvious considerable gains.

Cofinancing and Role of Commercial Banks

Cofinancing of development projects in the Third World is a rapidly changing activity the rise of which in recent years has been limited mainly to sources of a concessionary nature. However, a number of these sources have been encouraging cofinancing with private sources of funds, either through public issues by borrowing countries in the principal capital markets, through private placements by institutional investors, or by associating their financing with private bank loans.

Demand for capital from private institutions and commercial banks in particular will, in my view, be increasing in the future. The reason is relatively simple. The capital requirements for infrastructure and commercial investments have grown and will continue to grow — as a result of increasing economies of scale reflected in project design, as a result of the rapid technological change which often translates into higher capital, and last, but certainly not least, as a result of the ever-present inflation. In fact, the increase in capital requirements of individual projects is becoming such that many investment undertakings are escalating beyond the financing means of any single source. For instance, the two most recent projects approved for cofinancing by the Fund (the Arab Potash project in Jordan and the Guelb Iron Ore project in Mauritania) required each close to half a billion dollars in financing. Incidentally, the financial plans of both of these projects included substantial credits from commercial sources.

Looking into the future, the parade of projects where cofinancing is not only feasible but perhaps mandated by capital needs and shortages seems endless. The trend towards higher commodity prices despite the currently flagging economic situation, with recovery in industrialized countries yet to gather pace, will improve the export performance of developing nations. This, coupled with the current trend of tighter economic and fiscal policies in the Third World economies and greater discipline in their economic management in general, will hopefully bring about a decrease in lending risks and at the same time an increase in capital investments and requirements for economic development.

For these reasons, it would seem that project financing in developing countries by commercial banks offers significant potential. While the magnitude of external financing which may be or should be forthcoming from commercial banks can only be a subjective exercise at this time, there is no doubt that an acceptable level of external financing for developing nations, particularly those in the middle-income group, will only be achieved if commercial banks increase their exposure by substantial absolute amounts over the next several years.

On the macroeconomic level, the surplus from which all sources of external finance draw could well be directed towards new productive investments in the deficit countries, if only to avoid continuous inflation as a mechanism for maintaining global balance. In other words, to the extent that the savings realized in certain regions do not result from dissavings in other regions, they will either have to be matched up with new real investments or lead the way to further inflationary pressures.[6] There is increasing evidence, however, that under full international exchange the relatively advanced developing states will have a clear advantage as the place for many types of new investment. Large commercial banks in the West share with aid agencies the role of intermediation between surplus and deficit regions. Thus, in so far as they may succeed either directly or indirectly in the mobilization of surpluses for investment purposes in the developing world, they will be helping the strategies for maintaining a healthy world balance and achieving global development; there is also the probability of making a good profit in the meantime.

How, then, can the increase in commercial bank exposure come about without jeopardizing the financial position of these institutions in whose well-being we all share an obvious interest? In answering this question one must not lose sight of the special disadvantages of banks in terms of access to country information for purposes of risk assessment. Multilateral institutions, such as the IMF and the World Bank as well as the regional OPEC aid agencies — all with more ready access to country information — should do their part by making such data on indebtness and other critical performance indicators available in individual cases (with the consent of the governments concerned) and by encouraging exchanges of views between commercial bankers and concessionary aid donors. The same concessionary sources may also play a more active role in involving commercial sources, be they in Western or OPEC countries, in their activities, at least in the cases where cofinancing with the latter sources is likely to produce additional benefits to recipient nations.

Commercial and investment banks, on the other hand, could do their part

[6]The economist Hazem El Beblawi has established this point succinctly in his recent article, "Oil Financial Surpluses and the World Economic Structure," *Oil and Arab Cooperation*, vol. 4, no. 3 (1978) (in Arabic).

by introducing some flexibility into their lending terms. Experience has shown that the success of commercial projects is more sensitive to cash flows than to the cost of borrowing for such projects. If commercial banks could grant longer maturities, the viability of projects-particularly large, capital-intensive projects with a very gradual cash flow buildup -might be enhanced. This, of course, raises the issue of the short-term nature of many OPEC deposits which, some will argue, stands in the way of allowing the banks to go into long-term lending. Such an argument may be true, however, for a particular bank but cannot be true for the banking system as a whole. It should, therefore, be possible for commercial banks to harness collective action under the guidance, or, if preferred, with the cooperation of the central bank on the national level and with the IMF on the global level, to help protect the banks against the risks involved in the situation where they receive short-term deposits but extend long-term credits. Efforts should also continue towards creating more favorable objective conditions and more flexible banking techniques that would make longer-term deposits more attractive investment propositions.

In any event, the paramount benefit of cofinancing is the increase in external assistance it brings. Strictly from this perspective, changes in maturities and even in interest rates that could result from cofinancing may be of secondary importance. The OPEC Fund's experience, in a way unique since cofinancing has become the norm in its activities, has indicated certain areas in which cofinancing with commercial banks may be desirable if not, indeed, ideal. One example is the case of "green-field" commercial investments for which basic infrastructure support is required. In these projects, commercial investments and development aid can be pooled for the benefit of all. Under appropriate financial packages, a project may be structured into different components depending on the type of financing required for each. The commercial component can be financed with equity and commercial loans, while the infrastructure component can be financed with concessionary aid. The overall return on the total amount of capital employed will be such as to ensure a strong incentive for all parties on the basis of the project's merits alone. Other benefits of this scheme are obvious; the concessionary sources would benefit from the opportunity of diversifying their aid, both within a borrowing country, if they have definite allocations for each nation, and, more generally, within the geographical sphere of their lending activities. The commercial banks would still get a fair return on their investment. And the developing countries could benefit most from this marshaling of greater amounts of external capital, with the obvious advantages of increased investment, speedier and more efficient disbursements, and savings in time and procedures.

One must realize, nonetheless, that a blanket increase in financial cooperation among concessionary and private sources of finance is easier said than done. It has been mentioned earlier that commercial banks do face particular problems in increasing their exposure in developing countries. In Indeed, many of them already entertain the view that they may have overexposed themselves. The first difficulty is, of course, risk and, therefore, its appropriate assessment and trade-off with reward. This article has touched on the contribution that international aid institutions could play by disseminating, with the approval of governments, the information at their disposal. Another technique which could be applied more consistently is to include a cross-default provision in the loan agreements of concessionary cofinanciers which may enjoy a stronger security than commercial banks for their investments. One could assume, however, that such a technique would be used only to stimulate the flow of additional resources which would not be available otherwise on reasonable terms.

So far, other banks have tried to cover the additional risk of lending to developing countries with lower credit ratings mainly through applying higher interest rates, the higher spread being a self-insurance premium against default. One may ponder whether the benefits, in the terms of cost savings to consumers, which have resulted from the introduction of group plans in the insurance industry could be duplicated in the fields of banking. Specifically, could not the self-insurance sought individually by banks through their higher interest rate be transformed into a collective insurance pool, whereby risks would be uniformly spread in the industry and the cost to the borrowers reduced accordingly? It is hoped that collective self-interest would induce banks to work together for the establishment of such a scheme which probably wouid receive also the support of concessionary sources of finance. These in turn could consider the possibility of guaranteeing commercial credits, again substantially reducing the risks of commercial lenders. However, this latter type of cooperation must translate into net additional benefits to developing countries in the form of reduced cost of borrowing and incremental increases in capital transfer from commercial sources.

One should not unduly minimize the risks facing commercial banks. The risks in commercial bank lending are real, although in reality I think that the whole problem of less-developed countries' (LDCs') indebtedness is being magnified with the tendency to treat all developing nations as a single homogenous group. Some countries, no doubt, have relatively large outstanding debt burdens, and some may even already have borrowed beyond their prospective debt-servicing capacity. It is admittedly unfair, and, at any rate, unrealistic to request commercial banks to increase their exposure in the latter states. Such countries, unfortunately, may have to remain for a considerable time to come the favored clients of the most concessional sources of finance. Yet, just as a sneeze does not necessarily mean pneumonia, the problem of a few should not be interpreted as the widespread problem of all. It is possible to think at times that bankers are partly responsible for this tendency to generalize on the LDCs' debt problem. Nonetheless, when one reads that Chase Manhattan earned 65 percent of its profits abroad in 1977 and Citibank a staggering 82 percent, one cannot help but feel a bit uncomfortable with the perpetual single reference to the LDCs' debt problem as a constraint to more lending. These earning figures suggest, to this writer at least, that banks will maintain their outward-looking policy with regard to the developing world. The challenge now before them is whether they will be forward looking as well.

While project cofinancing between concessionary aid sources and commercial banks can be of mutual advantage to both types of institution, it does not offer the only appropriate channel for commercial banks to increase their project financing. Commercial banks are called upon to embark unilaterally as well as more aggressively in project financing even if it means diverting for this purpose part of their lending portfolio where the proceeds are generally not earmarked and, as a result, the productivity is, at best, questionable. In fact, banks and similar financing institutions could focus their lending on well-chosen self-liquidating projects where potential risks are certainly not greater than they would be in the case of general purpose lending. In addition to government guarantees for the loan, the tying of the repayment to project earnings provides further security. Some banks have built up impressive internal staff and capabilities to expand the finance of venture capital. Whether or not this will become a trend will depend on such factors as the initial success of banks gearing for this relatively new field of business and on the cooperation they provide and receive from larger multilateral aid agencies and from the borrowing countries themselves. The truth remains that an increase in project lending translates into greater real investment which, if related to sound priority projects, seems to offer, on a global basis, the solution for a world economy troubled with gross imbalances.

Conclusion

There has been much talk in the recent past about the interdependence of our world. International financial markets also have become sophisticated enough to forego the traditional classifications of "developed" and "developing" countries or "creditor" and "debtor" nations. International flows abound at present in all directions, at times bypassing the poor and helping the rich and at times leaving the former in the awkward position of financing the latter, as is often the case in domestic markets. It is indeed the responsibility of all sources of finance to see to it that adequate and enlightened management policies are developed to ensure not only the continuity and stability of such flows but their rational utilization as well.

Social justice and long-term stability dictate the imperatives of economic development in the less fortunate countries of the world, which still constitute the vast majority of nations. Just as interdependence among countries is real, the plight of the developing world *is* real and should not be considered as a distant fading echo that will eventually disappear through the work of unknown forces.

Greater access to capital is a complement, if not a prerequisite, to the goals of expanded trade that will fuel economic growth in both the developed and the developing countries. Sound and fair trade policies are invariably essential also. But foreign exchange earnings will contribute only a portion of the financing required for economic growth in developing countries — a growth that must take place and will take place only with continued access to external finance. Hence, the great need for closer cooperation between the two types of institutions involved — the agencies in charge of financial assistance and those in charge of capital markets.

12

OIL PRICES AND THE WORLD BALANCE OF PAYMENTS*

*John P. Powelson***

In assessing the impacts of oil price increases on industrialized and less-developed countries respectively, it is necessary to distinguish two separate effects: the financial and the real. The latter consists of the real amount of goods and services paid to the oil-producing countries because of the increased price of oil. The former consists of the money balances. Differences between the two occur, of course, because money balances are returned as oil-producing nations buy more goods and services. Thus payment in real terms mitigates payment in financial terms. In addition, oil-

---

* An earlier version of this chapter, covering data through 1975, was published in the *Journal of Energy and Development*, autumn 1977, under the title "The Oil Price Increase: Impacts on Industrialized and Less-Developed Countries."

** John P. Powelson is currently Professor of Economics and Research Associate, Institute of Behavioral Science at the University of Colorado (Boulder). Prior to joining the Colorado faculty, the author, who holds a Ph.D. in economics from Harvard University, had taught at the University of Pittsburgh, the Institute for International Development of the School of Advanced International Studies of Johns Hopkins University, the University of Buffalo, Harvard University, and the University of Pennsylvania's Wharton School of Finance and Commerce. His major nonacademic positions have included: Economist and Assistant Chief of Training, International Monetary Fund; Senior Economic Advisor, Ministry of Finance and Planning, Government of Kenya; Consultant to the Inter-American Development Bank (Washington, D.C.); Director, Program of Integrated National Accounts, Centro de Estudios Monetarios Latinoamericanos, Mexico City; and Economic Advisor to the Minister of Finance, Government of Bolivia. Professor Powelson has authored and edited some eight books, including *The Economics of Development and Distribution*, with William Loehr (New York: Harcourt, Brace, Jovanovich, 1981); *Economic Development, Poverty, and Income Distribution*, with William Loehr (Boulder, Colorado: Westview Press, 1977); *Employment in Africa*, with Philip Ndegwa (Geneva: International Labor Office, 1973); *Institutions of Economic Growth: A Theory of Conflict Management in Developing Countries* (Princeton, New Jersey: Princeton University Press, 1972); *Latin America: Today's Economic and Social Revolution* (New York: McGraw-Hill, 1964); *Economic Accounting* (New York: McGraw-Hill, 1974); and *National Income and Flow of Funds Analysis* (New York: McGraw-Hill, 1960). He has also written numerous articles, chapters in other volumes, and reviews. The author is indebted to Barbara Kuiper for assistance in preparing the tables in this study.

producing countries return financial balances through investment. Those they do not return may be kept (in their name) in banks of other countries, and it makes a good deal of difference in whose banks they are kept.

It would be convenient if we were to possess a consolidated regional balance-of-payments statement for OPEC members – consolidated to eliminate transactions among them and regional to show the flows by counterpart countries. Unfortunately, such a statement is not published, at least not in complete form. In the present article, therefore, I have estimated in very rough fashion, the differences between such a balance-of-payments statement for 1971 and for 1978. The word "rough" is emphasized, for the data are drawn from different sources, not all consistent with each other.

Even this exercise encounters conceptual problems. Balance-of-payments statements are published in current prices only. (A "real" statement would contain an item for the terms of trade.) Of course, we might estimate a "real" statement by taking into account what we do know of the terms of trade, but to do so would require detailed studies of commodity prices, which would delay our estimates beyond the date at which they would be useful. Readers will have to make their own guesses on price discounts for OPEC imports from both industrialized and less-developed countries.

Subject to this warning for "roughness," we can compare the balance-of-payments statements of the four groups shown in figures 1 and 2. Figure 1 compares 1975 with 1970, while figure 2 compares 1978 with 1971. The picture is somewhat as follows: in 1975, industrial countries paid $62.8 billion more for oil imports than they did in 1970; the 1978/1971 difference is $84.4 billion. OPEC returned to them $34.9 billion of this excess in 1975 by buying more goods and services, an amount that increased to $72.9 billion in the 1978/1971 comparison. Thus, OPEC purchases from the industrialized world were creeping up, tending to alleviate the balance of payments. The 1975/1970 comparison shows a net flow of investments *out* ($4.2 billion) from OPEC to industrialized countries, but by 1978 this net flow had been reversed and there was a net flow *in* ($1.8 billion) for the 1978/1971 comparison.

With the non-oil-exporting less-developed countries (LDCs), however, the picture is one of significant increases in oil payments ($11.4 billion, 1975/1970, and $24.1 billion, 1978/1971), with lesser increases of OPEC imports from these countries ($5.2 million, 1975/1970, and $9.4 million, 1978/1971). Aid from OPEC to the non-oil-exporting LDCs has been negligible compared to these amounts.

The other impact – that on balance of payments – depends on how it is counted. Reserves of about $16.3 billion (1975 increment over 1970) flowed from both industrialized and less-developed countries to OPEC members. Virtually all of these reserves, however, were maintained in the banks (or short-term securities) of the industrialized countries. Hence, the impact on

Figure 1

SCHEMATIC BALANCES OF PAYMENTS OF FOUR COUNTRY GROUPS, EXCESS OF 1975 OVER 1970
(billions of dollars)

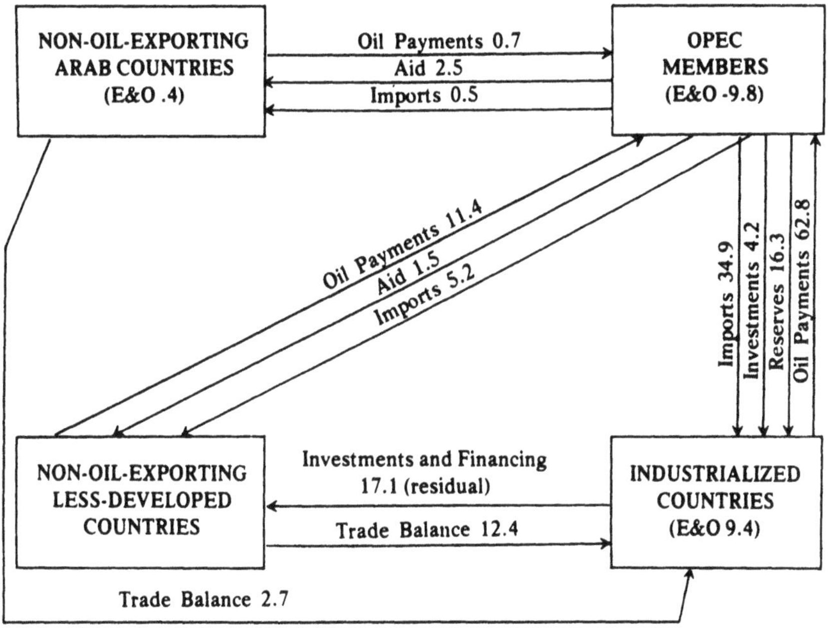

Arrows show direction of *payments;* imported goods flow in opposite direction.

Sources: International Monetary Fund, *International Financial Statistics* and *Direction of Trade*, various issues; additional data from Organization for Economic Cooperation and Development (OECD), *Development Cooperation, 1976 Review* (Paris: OECD, 1976), p. 117.

their money supplies was nil or minimal. As mentioned earlier, it makes a great difference in which countries the reserves are kept. While I have done no analysis of this, it would appear that the OPEC finance ministers have been responsible, in that they have not caused serious currency dislocations, which it would have been within their power to do.

From 1975 on, the picture becomes less clear. Instead of a universal flow of reserves into OPEC, we find those countries splitting into two groups. For the Middle Eastern Arab countries, the current account is still highly positive. Only the data on Kuwait appear in table 1, however, for Saudi Arabia and Iraq are late in reporting. But Algeria, Ecuador, Indonesia,

Figure 2

SCHEMATIC BALANCES OF PAYMENTS OF FOUR COUNTRY GROUPS, EXCESS OF 1978 OVER 1971
(billions of dollars)

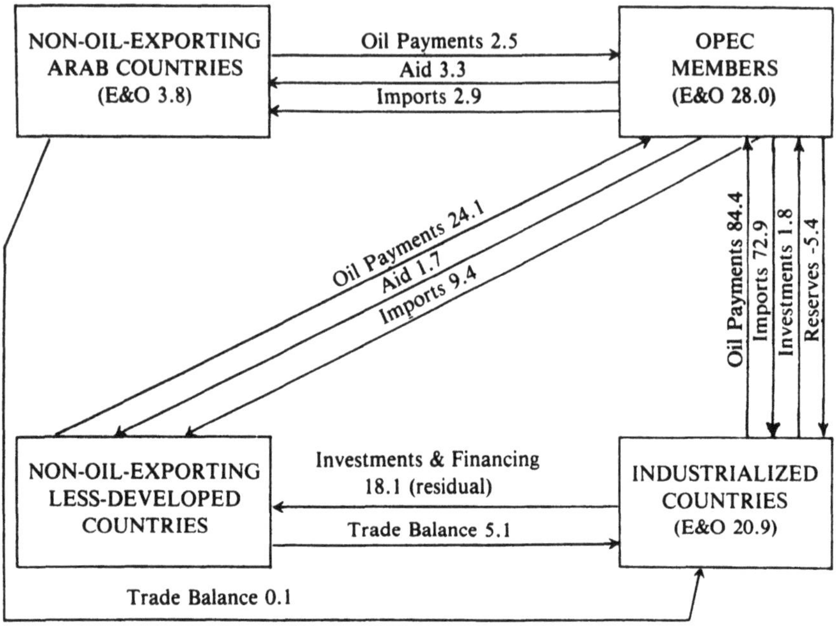

Arrows show direction of *payments*; imported goods flow in opposite direction.

Source: Tables 3 and 4. Investments and financing (industrialized to nonoil LDCs) is here calculated as residual for nonoil LDCs (hence zero errors and omissions), but it roughly corresponds to amounts shown in table 2. Errors and omissions (E&O) for all other areas are residuals, representing transactions with third countries and international institutions, as well as E&O in the original balance-of-payments statements.

Nigeria, and Venezuela have so stepped up their imports that they run increasing deficits on current account (table 1). By the same token, Kuwait has become an exporter of long-term investment, while the second group has become heavy importers of long-term capital. Likewise, while the first group is accumulating reserves, the second has been drawing them down. As a result, the 1978/1971 comparison for OPEC as a whole shows the net reserve accumulation as decreasing ($5.4 billion, figure 2).

The situation remains increasingly grave for the non-oil-exporting LDCs. They paid $11.4 billion more for their purchases of oil in 1975 than in 1970,

Table 1

CONDENSED BALANCE OF PAYMENTS OF OPEC COUNTRIES,
1971, 1974, 1976, 1978
(in millions of SDRs)

	1971	1974	1976	1978
1. Algeria				
Exports of oil and gas	657	3,351	4,065	4,856
Government grants (net)	196	-354	-12	-10
Other current items (net)	-810	-2,865	-4,820	-7,762
Balance on current account	43	132	-767	-2,826
Net capital inflow	85	299	1,323	2,960
Net increase in reserves	128	431	556	134
2. Ecuador				
Exports of crude petroleum	1	659	637	446
Government grants (net)	8	13	18	35
Other current items (net)	-165	-641	-661	-601
Balance on current account	-156	31	-6	-120
Net capital inflow	145	60	182	125
Net increase in reserves	-11	91	176	5
3. Gabon				
Exports of oil products	98	589	783	735
Government grants (net)	8	16	37	30
Other current items (net)	-88	-431	-796	-674
Balance on current account	18	174	24	91
Net capital inflow	-9	-132	-39	-85
Net increase in reserves	9	42	-15	6
4. Indonesia				
Exports of oil sector	515	4,212	5,267	5,879
Government grants (net)	46	41	13	11
Other current items (net)	-932	-3,756	-6,068	-6,900
Balance on current account	-371	497	-788	-1,010
Net capital inflow	370	75	1,570	1,131
Net increase in reserves	-1	572	782	121
5. Iran				
Exports of oil sector	3,354	17,082	20,003	--
Government grants (net)	4	-27	-17	--
Other current items (net)	-3,476	-6,855	-15,903	--
Balance on current account	-118	10,200	4,083	--
Net capital inflow	506	-4,358	-3,702	--
Net increase in reserves	388	5,842	381	

Table 1

CONDENSED BALANCE OF PAYMENTS OF OPEC COUNTRIES,
1971, 1974, 1976, 1978 (continued)
(in millions of SDRs)

	1971	1974	1976	1978
6. Iraq				
Exports of oil sector	1,479	5,707	7,44	8,329
Government grants (net)	1	-197	--	--
Other current items (net)	-1,288	-3,322	--	--
Balance on current account . . .	192	2,178	--	--
Net capital inflow	-101	-588	--	--
Net increase in reserves	91	1,590	1,613	--
7. Kuwait				
Exports of oil sector	2,136	7,683	7,696	7,515
Government grants (net)	--	--	-193	-639
Other current items (net)	--	--	-1,485	-1,951
Balance on current account . . .	--	--	6,018	4,925
Net capital inflow	--	--	-3,239	-3,012
Net increase in reserves	38	728	2,779	1,913
8. Libya				
Exports of oil sector	2,701	5,992	7,568	7,864
Government grants (net)	-90	-57	-125	-218
Other current items (net)	-1,830	-4,412	-5,334	-6,828
Balance on current account . . .	781	1,523	2,109	818
Net capital inflow	84	-334	-1,223	-1,608
Net increase in reserves	865	1,189	886	-790
9. Nigeria				
Exports of oil sector	1,373	7,489	8,180	7,493
Government grants (net)	27	-6	5	-16
Other current items (net)	-1,805	-3,411	-8,494	-10,489
Balance on current account . . .	-405	4,072	-309	-3,012
Net capital inflow	579	46	-3	979
Net increase in reserves	174	4,118	-312	-2,033
10. Saudi Arabia				
Exports on oil sector	3,488	24,990	30,622	--
Government grants (net)	-68	-844	-2,883	--
Other current items (net)	-2,513	-5,016	-15,788	--
Balance on current account . . .	907	19,130	11,951	--
Net capital inflow	-237	-10,679	-8,609	--
Net increase in reserves	670	8,451	3,342	--

Table 1

CONDENSED BALANCE OF PAYMENTS OF OPEC COUNTRIES,
1971, 1974, 1976, 1978 (continued)
(in millions of SDRs)

	1971	1974	1976	1978
11. Venezuela				
Exports of oil sector	2,905	8,785	7,624	6,953
Government grants (net)	-4	-53	-54	-30
Other current items (net)	-2,914	-3,901	-7,203	-11,210
Balance on current account	-13	4,831	367	-4,287
Net capital inflow	467	-1,115	-345	2,774
Net increase in reserves	454	3,716	22	-1,513
Total except Iraq and Kuwait				
Exports of oil sector	15,092	73,149	84,849	--
Government grants (net)	127	-1,271	-3,018	--
Other current items (net)	-14,533	-31,288	-65,067	--
Balance on current account	686	40,590	16,664	--
Net capital inflow	1,990	-16,138	-10,846	--
Net increase in reserves	2,676	24,452	5,818	--
Total except Iraq, Kuwait, and Saudi Arabia				
Exports of oil sector	11,604	48,159	54,127	--
Government grants (net)	195	-427	-135	--
Other current items (net)	-12,020	-26,272	-49,279	--
Balance on current account	-221	21,460	4,713	--
Net capital inflow	2,227	-5,459	-2,237	--
Net increase in reserves	2,006	16,001	2,476	--
Total				
Exports of oil sector	18,707	86,539	100,289	--
Government grants (net)	--	--	--	--
Other current items (net)	--	--	--	--
Balance on current account	--	--	--	--
Net capital inflow	--	--	--	--
Net increase in reserves	2,805	26,770	10,210	--

Source: International Monetary Fund, *Balance-of-Payments Yearbook*, vol. 30 (1979).

and they received $5.2 billion in exchange for their exported products. The latter would constitute the "real" loss to them, though of course it is mitigated by increases in the prices of some primary product exports (such as coffee and sugar). Even so, the real price increase for them, as a percentage of their (much lower) gross national products (GNPs), was far more severe than it was for the industrialized countries.

The impact on nonoil LDCs becomes more severe when indirect effects are counted. Not all LDCs have their own refineries, so to a large extent their purchases of petroleum products are not made directly from OPEC. For example, petrochemical products (e.g., fertilizers) are supplied largely by the industrialized nations. Indirectly, therefore, part of the $62.8 billion paid by the industrialized countries covered oil whose increased final consumption occurred in LDCs. I have not estimated how much this part would be. However, the trade balances of nonoil LDCs with industrialized nations deteriorated by $12.4 billion (1975 compared with 1970).

Since 1975, the debt of non-oil-exporting LDCs has accelerated. While I would not expect my figures to correspond exactly (because of different sources), nevertheless the "investments and financing" (residual flow) given for 1975 in figure 1 ($17.1 billion) does correspond roughly with the World Bank data on disbursed debt in that year ($20.1 billion, table 2). For 1978, however, the comparison diverges greatly. Our trade-and-aid residual in figure 2 is $18.1 billion for the 1978/1971 comparison, not much increased from the 1975/1970 comparison, whereas the World Bank shows nonoil LDCs as increasing their debt in 1978 by some $40.1 billion (table 2).

Foreign aid and other resource flows from OPEC to nonoil LDCs have principally gone to members of the Arab League (see table 3); in 1978 they

Table 2

DISBURSED DEBT OUTSTANDING OF 87
NON-OIL-EXPORTING LESS-DEVELOPED COUNTRIES
(in millions of U.S. dollars)

Year	Outstanding	Annual Increment
1972	62,067	--
1973	74,110	12,043
1974	92,497	18,378
1975	112,648	20,151
1976	138,194	25,546
1977	169,946	31,752
1978	210,013	40,067

Source: World Bank, *World Debt Tables*, vol. 1, (EC-167/79), December 28, 1979, table 1-A, p. 18.

amounted to $3.3 billion. Of the $1.7 billion to other LDCs, about $0.2 billion went to India, about $0.1 billion to Pakistan, and the rest scattered in smaller amounts to other countries.

The deficit of nonoil, non-Arab League LDCs has, of course, been financed largely by loans from governments and banks in the industrialized world. In the mid-'70s, the mainstream of economic opinion was that these debts were not excessive, for the countries concerned could service them out of growing exports. But recession in the industrialized world has dampened this optimism. More and more, banks in industrialized countries tend to refuse further loans and to harden the terms on those they permit. It is now estimated that Brazil faces a payments gap of $11 billion in 1980; Argentina $2.7 billion; Turkey $2.5 billion; Thailand $2.4 billion; the Philippines $2.0 billion; and South Korea $1.7 billion (*New York Times,* April 14, 1980).

The Balance of Payments

Balance-of-payments statements are published by the International Monetary Fund in both *International Financial Statistics* (IFS) and the *Balance-of-Payments Yearbook* (BPY). Normally, the IFS data are more up-to-date, but for the present study it was just as convenient to use BPY. Table 1 presents a breakdown of information for the individual countries of OPEC. It does not tie into the other tables, however, because not all of OPEC has prepared balance-of-payments tables for recent years. Hence

Table 3

TOTAL OFFICIAL FLOW OF RESOURCES TO DEVELOPING COUNTRIES FROM OPEC AND ARAB/OPEC MULTILATERAL INSTITUTIONS, 1975-78
(in millions of U.S. dollars)

					Percentage of Total	
	1975	1976	1977	1978	1975	1978
Arab League members	5,229.8	4,332.2	4,297.6	3,295.7	78.5	65.7
India	203.7	499.6	185.6	197.9	3.0	4.0
Pakistan	458.4	840.8	73.8	117.6	6.9	2.3
American countries	207.7	221.8	374.6	139.1	3.1	2.8
Other Asia	170.5	216.3	352.5	205.8	2.6	4.1
Other Africa	290.1	186.2	157.7	245.3	4.4	4.9
Other	99.2	402.6	840.7	814.1	1.5	16.2
Total	6,659.5	6,699.5	6,282.4	5,015.5	100.0	100.0

Source: Organization for Economic Cooperation and Development (OECD), *Development Cooperation, 1979 Review* (Paris: OECD, 1979), pp. 278-79.

there will be some discrepancies between these data and those of tables 2, 3, and 4. Table 1 data are in special drawing rights (SDRs), the other tables in U.S. dollars.

OPEC Country Trade

In the final analysis, the increased price of oil is paid or not paid in real goods and services. In order to understand fully the relative impacts on industrialized and less-developed countries, it would be necessary to disaggregate imports and exports (including oil) according to commodity groups, and to separate out the price and quantity changes. The operation would be tedious, and the detailed data have not yet been assembled by international institutions. We therefore must be content with information compiled by the International Monetary Fund in *Direction of Trade* (DOT).

Table 4 is mainly for the record, because it is the basis for the trade flows

Table 4

EXPORTS AND IMPORTS OF OIL-EXPORTING COUNTRIES, 1971, 1978, BY AREA
(in millions of U.S. dollars)

	1971		1978		Change		Percentage Increase
	Amount	%	Amount	%	Amount	%	1978/71
Exports to:							
Industrialized countries	16,080	75.1	100,472	71.4	84,392	70.7	525
Arab League countries	377	1.8	2,885	2.0	2,508	2.1	665
Other nonoil developing countries	3,245	15.2	27,337	19.4	24,092	20.2	744
Rest of world	1,697	7.9	10,105	7.2	8,408	7.0	496
Total	21,339	100.0	140,799	100.0	119,400	100.0	558
Imports from:							
Industrialized countries	8,952	78.0	81,803	79.1	72,851	79.3	814
Arab League countries	384	3.3	3,018	2.9	2,634	2.9	686
Other nonoil developing countries	919	8.0	10,297	10.0	9,378	10.2	1,021
Rest of world	1,225	10.7	8,226	8.0	7,001	7.6	572
Total	11,480	100.0	103,344	100.0	91,864	100.0	800

Source: International Monetary Fund, *Direction of Trade*, Annual 1972-78 and 1971-72.

in figure 2. It shows that, to finance their increased purchases of oil, nonoil LDCs had to increase their exports to OPEC some tenfold (from $919 million in 1971 to $10,297 million in 1978), while the industrialized countries increased theirs eightfold, from $8,952 million to $81,803 million. In the 1975/1970 comparison, the trade deficit of the industrialized countries increased by 755.9 percent, compared with 607.5 percent for the nonoil LDCs. But in the 1978/1971 comparison, the industrialized countries' deficit was increased by only 162 percent, compared with 532.6 percent for the nonoil LDCs. This change reflects the increased purchases of OPEC in the industrialized world relative to the nonoil LDCs.

Finally, the real burden should not be equated to the goods and services in any single current year. Rather, it is measured (for a given year) by the sum of three items: increased exports, increased debt, and reduction in reserves. The increased exports are the portion of the real burden that a country (or group of countries) pays in the current year; the increased debt represents the real payments it must make in subsequent years for the increased oil price of this year; and the decreased reserves represent purchasing power foregone.

The Debt Problem

The increased exports that a country must yield (loss in the terms of trade) are burden enough, but at least they are paid in the current year. It is the unpaid portion of current oil imports that hangs over the countries' heads for years to come. This problem faces both industrialized and nonoil LDCs, but in different ways. The increased reserves held by OPEC members in the industrialized countries constitute debt by the latter to the former, in the same sense that a currency note is debt by the central bank to the holder. Reserves are a claim against future goods and services to be produced by the industrialized countries. There are, however, various ways by which industrialized nations can defend themselves in case the claim is exercised, inflation and devaluation being not the least. For the less-developed countries, whose debts are specified in currencies other than their own, the threat of bankruptcy is greater and is less within their control.

Table 2 presented the World Bank data on debt of the nonoil LDCs. Not only has the amount outstanding increased each year from 1972 to 1978, but the yearly increment has also increased. Two factors mitigate the severity of the debt burden. One is world inflation. The outstanding debt of $210,013 million in 1978 would have been only $134,637 million 1972 dollars (deflated by the consumer price index in the United States). In real terms, therefore, it increased by 117 percent from 1972 to 1978.

The other factor is the annual increase in the GNPs of the developing countries, which improves their capability to service and repay debt. Here, too, the problem is worsening (see table 5). The ratio of disbursed debt outstanding to GNP for low-income countries increased from 20.6 percent in 1974 to 25.6 percent in 1978 and for middle-income countries rose from 15.4 percent to 23.2 percent. The ratios of disbursed debt outstanding, of debt service, and of interest payments to exports also increased for both groups of countries (table 5).

The severity of the debt burden is normally judged through two ratios: the time profile and the debt service. The former is the ratio of debt payments (both principal and interest) for a given number of years (say, five or 10) to the amount outstanding at a given date. The latter is the ratio of debt service (both principal and interest) to exports of goods and nonfactor services for the same time period, usually one year. The time profile ratios for all developing countries at the end of 1969 and the end of 1974 are shown in table 6, and the debt service ratios of selected developing countries are in table 7.

The time profile ratios reflect a hardening in terms between the two years (the five-year ratio increasing from 48 to 52, the 10-year from 84 to 90, for 86

Table 5

DEBT AND DEBT SERVICE INDICATORS
(percentages)

	1974	1977	1978
Disbursed debt outstanding/GNP			
Low-income countries	20.6	25.4	25.6
Middle-income countries	15.4	20.1	23.2[a]
Disbursed debt outstanding/Exports			
Low-income countries	147.1	156.1	168.5
Middle-income countries	59.0	81.8	94.1
Debt service/Exports			
Low-income countries	10.9	13.2	15.2
Middle-income countries	10.5	13.0	17.1
Interest payments/Exports			
Low-income countries	3.2	4.6	5.2
Middle-income countries	3.4	4.3	5.3

Source: World Bank, *World Debt Tables*, vol. 1 (EC-167/79), December 28, 1979, table IV, p. 5.

[a]The 1978 middle income country debt-to-GNP ratio excludes Iran, for which GNP data were not available. The 1978 exports were estimated on the basis of merchandise exports, with nonfactor services and workers' remittances included in the same proportion as in 1977.

developing countries). These ratios reflect both an increase in interest rates and increased reliance on private sources, whose terms are almost always harder than those of governments and international agencies.

Debt service ratios also increased between the years compared. In assessing them, one should not discount for the mitigating factors mentioned above, for both parts of the ratio are expressed in nominal values – hence the inflation mitigator does not apply – and exports ought to increase with gross domestic product (GDP) – hence the growth mitigator does not apply. In other words, the erosion by inflation and growth, considered above, would cause the debt service ratio to fall. The fact that it is instead rising is cause for concern.

Table 6

DEBT SERVICE TIME PROFILE RATIOS FOR DEVELOPING COUNTRIES, 1969 and 1974

	1969		1974	
	5-Year Ratio	10-Year Ratio	5-Year Ratio	10-Year Ratio
1. 86 developing countries	48	84	52	90
Loans from governments	39	71	38	71
Loans from international organizations	38	75	31	69
Loans from private sources	75	106	79	124
2. Selected regions				
Latin America and Caribbean:	58	91	67	85
Loans from governments	43	72	50	85
Loans from international organizations	45	91	43	90
Loans from private sources	82	109	85	132
South Asia:	34	63	30	55
Loans from governments	30	61	31	60
Loans from international organizations	26	50	16	31
Loans from private sources	78	116	78	106

Source: World Bank, *World Debt Tables*, vol. 1 (EC-167/76), October 31, 1976, p. 18.

Table 7

DEBT SERVICE RATIOS FOR SELECTED DEVELOPING COUNTRIES
(as a percentage of exports of goods and nonfactor services)

	1967	1970	1973	1974
Ratios over 20% in 1973				
Egypt[a]	19.4	26.2	35.0	32.0
Mexico[c]	24.6	25.2	25.2	18.4
Pakistan[a]	18.5	27.8	21.9	15.3
Peru[a,c]	11.1	13.9	32.2[b]	25.6
Uruguay	17.0	19.2	29.2	21.8
Zambia	2.4	5.7	28.5	5.1
Ratios 15-20% in 1973				
Afghanistan[a]	11.2	19.8	19.5	17.9
Algeria	2.0	3.7	15.1	14.4
Argentina[a,c]	25.9	21.0	17.8	16.2
Burma	6.4	16.0	18.7	14.4
India[a]	22.5	22.3	18.0	15.9
Iran	5.0	11.4	15.9[b]	6.8
Israel	15.7	18.6	17.7	18.7
Nicaragua	6.4	10.6	17.8	10.7
Ratios 10-15% in 1973				
Bolivia	5.9	10.9	15.0	11.6
Brazil[a,c]	15.6	15.1	13.1	15.2
Chile[a]	12.6	18.3	11.8	11.1
Colombia	14.2	11.9	13.6	16.7
Costa Rica	12.0	9.7	10.2	9.4
Gabon	4.7	5.4	14.7	4.4
Korea	6.2	21.0	10.3	10.5
Paraguay	7.4	11.2	10.3	7.8
Sri Lanka	3.9	9.7	12.9	11.2
Sudan	5.6	9.2	10.9	11.7
Tunisia	20.6	19.5	12.0	7.2
Turkey[a]	16.4	-2.5	13.1	12.2
Ratios below 10% in 1973 but above 10% in 1974				
Somalia	2.1	2.1	3.6	13.4
Zaire	2.4	4.6	8.5	11.7

Source: World Bank, *World Debt Tables*, vol. 1 (EC-167/76), October 31, 1976, p. 19.

[a]Debt service payments reduced in some years shown through debt relief arrangements.
[b]Debt service figures include prepayments.
[c]These ratios do not include service on private sector debt which is significant for the countries indicated.

Some 15 years ago, economists generally believed that debt service ratios in the neighborhood of 10 to 15 percent were the highest a country ought to incur. Beyond that point, its credit rating might be impaired. As the ratios for many LDCs have advanced and general bankruptcy has not occurred, the tolerable limit itself has risen. It is not that countries are any better able to service their debts now than they were before; rather the general belief is that earlier assessments were too conservative.

The World Bank has not published debt service ratios or time profiles in its latest reports. Instead, we have data on the average terms of loan commitments of official and private creditors, 1972-1978 (see table 8). These data reflect further tightening, in that interest rates are rising (in accordance with worldwide trends), while the grant element of loans is decreasing (except for international organizations). While average maturities were decreasing through 1977, they appear to have been relaxed a bit in 1978, bringing them close to par with 1972.

The Expanded Assistance Program

There should, therefore, be no room for complacency; as long as countries continue to borrow to meet the high prices of petroleum and for other purposes, both the outstanding debt and debt service ratios will continue to rise. To help meet this problem, OPEC members announced an expanded assistance program in 1976. A multitude of agencies had already existed, so much so that OPEC and Arab assistance funds have been criticized for their dispersion and inefficiency.[1] In January 1976, the OPEC Special Fund was formed with initial resources of $800 million; it has since been increased to $1.6 billion (August 4, 1977). The Fund will operate through three separate programs. First is the International Fund for Agricultural Development (IFAD), to which member countries of the Organization for Economic Cooperation and Development (OECD) have also agreed to contribute ($1.50 for every $1 by the OPEC Fund). The second is the balance-of-payments support program for LDCs. By the end of 1976, $197 million had been committed to 49 developing countries in Asia, Africa, and Latin America, of which only three were members of the Arab League (Egypt and the two Yemens), and their share was less than 10 percent. This fund therefore may represent a breakthrough in support to non-Arab League members. The third channel will be the Trust Fund administered by the IMF. In October 1976, the OPEC ministers of finance recommended that member

[1] Ragaei El Mallakh and Mihssen Kadhim, "Arab Institutionalized Development Aid: An Evaluation," *Middle East Journal*, autumn 1976, p. 479.

Table 8

AVERAGE TERMS OF LOAN COMMITMENTS OF OFFICIAL AND PRIVATE CREDITORS, 1972-1978

	1972	1975	1977	1978
Total official lenders				
Amount (billions of $)	11.6	21.9	25.8	27.2
Interest rate (%)	4.3	4.9	5.2	5.0
Maturity (years)	24.2	23.5	22.6	24.8
Grant element (%)	39.9	36.1	33.9	37.0
Government lenders				
Amount (billions of $)	7.9	13.8	13.5	13.4
Interest rate (%)	3.8	4.2	4.6	4.7
Maturity (years)	22.7	21.1	22.5	23.9
Grant element (%)	42.6	39.2	38.0	38.7
International organizations				
Amount (billions of $)	3.8	8.1	12.3	13.8
Interest rate (%)	5.5	6.1	5.9	5.2
Maturity (years)	27.5	27.7	22.9	25.7
Grant element (%)	34.2	31.0	29.4	35.4
Total private lenders				
Amount (billions of $)	8.8	23.8	37.0	49.8
Interest rate (%)	7.3	8.8	8.0	9.4
Maturity (years)	8.9	7.8	8.0	8.9
Grant element (%)	10.9	4.7	7.8	1.7
Suppliers				
Amount (billions of $)	2.4	5.5	7.7	6.1
Interest rate (%)	7.0	7.9	7.8	7.9
Maturity (years)	9.4	9.9	11.1	10.3
Grant element (%)	11.4	8.5	9.0	7.5
Total financial markets				
Amount (billions of $)	6.4	18.3	29.3	43.8
Interest rate (%)	7.4	9.0	8.1	9.6
Maturity (years)	8.7	7.1	7.2	8.8
Grant element (%)	10.7	3.6	6.5	0.9
Financial institutions				
Amount (billions of $)	5.5	17.4	25.6	40.0
Interest rate (%)	7.4	9.1	8.1	9.8
Maturity (years)	7.7	7.0	7.1	8.7
Grant element (%)	9.2	3.0	6.0	-0.4
Bonds				
Amount (billions of $)	0.9	0.9	3.8	3.8
Interest rate (%)	6.9	8.0	7.8	7.2
Maturity (years)	14.5	10.1	7.9	9.3
Grant element (%)	19.8	13.6	9.7	14.4

Source: World Bank, *World Debt Tables*, vol. 1 (EC-167/79), December 28, 1979, p. 194.

countries contribute to the Fund their portions of the profits from sales of Fund gold held in their names. Other non-OPEC countries are expected to do the same. This channel is therefore "free" in the sense that it does not require contributions from current resources.

At the Arab-African summit conference in March 1977, Saudi Arabia pledged an additional $1 billion in aid to Africa; Kuwait, Qatar, and the United Arab Emirates followed with pledges of $450 million. It has not yet been announced exactly how these funds are to be allocated or when.[2] Table 3 showed that the total flow of OPEC resources decreased rather than increased from 1977 to 1978.

Conclusion

The impact of increased oil prices is far-reaching and is so intertwined with other economic events that it is difficult to determine the extent to which it has affected less-developed as opposed to industrialized countries. But some broad generalizations can be made. First, the industrialized countries have been able to afford the higher payments in real terms, and their financial positions have been spared because of increased OPEC purchases from them, investments in them, and holding of reserves in their bank accounts or other financial instruments.

The non-oil-exporting LDCs have not been so fortunate. They have paid more in real terms for oil, both in direct payments to OPEC members and indirectly through the purchase, at higher prices, of petroleum products from industrialized countries. These real payments have lessened their abilities to carry out development programs. Because OPEC members have not, in general, invested in less-developed countries, nor kept their reserves in LDC currencies, the financial condition of LDCs is more precarious each year. Their deficits have been financed partly by international lending institutions and increasingly by banks in industrialized countries. Both these sources of financing are reaching their limits, and there is some indication that each would like to shunt the problem to the other. In connection with the New International Economic Order, suggestions for debt moratoriums or debt forgiveness are increasingly heard. Neither is likely to occur, except to stave off immediate bankruptcies, because of a strongly held feeling in the industrialized countries that debt forgiveness should not be a principal instrument of foreign aid.

The increased lending programs of OPEC members are a refreshing

---

[2] *Middle East Economic Survey,* March 14, 1977, p. 9.

breath, albeit small, that may postpone the crisis. But even the expanded amounts involved are far from what would be necessary to restore the *status quo ante* for nonoil LDCs; clearly, they are not a permanent solution. The "oil glut" of 1981 and the refusal of Saudi Arabia to countenance an increase in OPEC prices in May of that year may lend a moratorium to *accelerated* debt increases, but unless LDCs reduce their consumption of oil, debt will still go up. As long as debts are rising, the problem will not go away. It remains perhaps the principal crisis on the international economic scene with which the world has not yet come to grips.

13

FRIENDS OR FELLOW TRAVELERS? THE RELATIONSHIP OF NON-OPEC EXPORTERS WITH OPEC

*Øystein Noreng**

Ambiguous Outsiders

In a historical perspective, the success of OPEC and the ability of the OPEC member countries to assert their own interests are salient features of the 1970s, representing a shift in international power relations on a crucial dimension in favor of a group of raw material exporters that also are developing countries. This has had a global impact on industrialized consumers of petroleum as well as oil-importing developing nations. It has affected in particular a small number of countries which are net exporters of petroleum, but which are outside OPEC and which in general have more diverse economic interests than those related to petroleum alone. As a function of OPEC's success, the non-OPEC exporters receive a much higher price than previously for their petroleum, and their bargaining position with both the international oil industry and the importing nations has been improved significantly. However, because of the diversity of their economic interests, their petroleum interests should be seen within a wider context.

Some of the non-OPEC exporters are rich, industrialized countries, sharing important interests with the industrialized consumers of petroleum; others are poorer developing states, sharing important interests with others in the developing bloc. Consequently, these outsiders to OPEC could have an ambiguous relationship to OPEC's success to the extent that their oil

*Øystein Noreng, currently Professor at the Oslo Institute of Business Administration, Norway, earned a doctoral degree from the Sorbonne (Paris). His earlier positions included: Research Fellow, Norwegian Council for Scientific Research; Counselor, Planning Department, Royal Ministry of Finance; and Research and Planning Manager, Marketing Department, Statoil (the Norwegian national oil company). Dr. Noreng's book as part of the series on the 1980s Project of the Council on Foreign Relations was published in 1978 as *Oil Politics in the 1980s: Patterns of International Cooperation* (New York: McGraw-Hill Book Company). His volume on *The Oil Industry and Government Strategy in the North Sea* is forthcoming in mid-1979 (Boulder, Colorado: International Research Center for Energy and Economic Development).

interests and their other economic interests may be seen as contradictory. Their conscious relationship with OPEC may then depend upon an explicit trade-off between their petroleum interests and their other economic interests. To the degree that such an explicit trade-off is difficult, the non-OPEC exporters may prefer to enjoy more or less discreetly the fruits of OPEC's success, rather than to openly support OPEC. Thus, the position of a fellow traveler may appear politically as more comfortable than that of declared friendship.

However, such reasoning is essentially static. The degree to which the petroleum interests of the non-OPEC exporters can be seen as contradictory to their other economic interests essentially depends upon the time horizon considered. The non-OPEC exporters and the OPEC countries essentially face the same basic question in petroleum policy: to produce or not to produce? Both sets of petroleum exporters have a common responsibility concerning the management of the long-term global energy balance, i.e., to make demand trends and supply trends match by stimulating conservation and the development of new sources of energy. While this is a matter of worldwide concern and a responsibility also of the industrialized and developing consumers, in the present situation and with depletable petroleum as the world's major energy source, the oil exporters have a particular responsibility. Consequently, the non-OPEC exporters and the OPEC countries share important long-term interests; the relationship between them also should be seen in a dynamic perspective.

Presently, the member countries of OPEC account for approximately 80 to 85 percent of the *oil* entering international trade. Their exports of *natural gas* represent only a minor part of world trade in that commodity in spite of their large gas reserves. The remainder of the oil entering international trade is furnished by the Soviet Union, Mexico, Norway, China, Malaysia, Egypt, Oman, Trinidad, Syria, and Angola. Additionally, Great Britain is likely to become a net exporter of oil for some time during the 1980s. Several developing countries might become net oil exporters in the 1980s as well. The important exporters of natural gas are essentially non-OPEC countries, i.e., the Soviet Union, Canada, the Netherlands, and Norway; in addition Mexico is likely to become a major exporter of this energy source within a few years. Thus, the non-OPEC petroleum exporters make up a diverse group of countries: some, such as Canada, the Netherlands, and Norway, are rich, industrialized countries; some, such as Mexico, Malaysia, Oman, and Syria, are developing countries; some are considered major powers, such as the Soviet Union and China.

The essential common denominator among these countries, and between these nations and the OPEC members, is that they are net exporters of oil and have economic interests in relation to petroleum, which may be more or less important in relation to their other economic interests. As a rule, their standing in the world and their level of economic and social development are not overwhelmingly a function of their petroleum industry. The oil sector generally does not represent a major part of the gross national product, except in some marginal cases such as Oman and Trinidad. Petroleum revenues are in all cases important and improve generally the freedom of action in economic policy, but they can also create problems of adaptation and transition, create inflationary pressures, have a negative impact upon overall industrial competitivity, and thus to some extent substitute other income rather than improving the general economic performance. This seems to be particularly relevant for industrialized countries with substantial exports of petroleum, such as the Netherlands and Norway; it could also be relevant for Canada and Mexico, in case of quickly rising oil exports.

By contrast, for the OPEC countries their level of economic and social development, their terms of trade with the rest of the world, and their political standing internationally are linked essentially to the petroleum industry. As a rule, the petroleum sector here represents the major part of the gross national product, and the oil-generated revenues are crucial to both the public budget and to the financing of economic development plans. Thus, OPEC's success or failure is largely their own success or failure.

It should be recalled that the OPEC states generally have a colonial or semi-colonial past, having been subject to foreign economic and political domination. For these countries, the 1970s constitutes an important step in the direction of national and political emancipation through the price rise and the nationalization of the oil industry.[1] In the future, their economic development, terms of trade, and standing in the world essentially will depend upon OPEC's performance for a substantial period of time. Thus it is critical for them to maintain political cohesion and solidarity in spite of obvious differences in interests and points of view. This explains the ideological cohesion of OPEC, which contributes to its chances of survival and makes minor price disputes no immediate threat to unity.[2]

The non-OPEC exporters do not share the historical success of OPEC in the same way for in spite of obvious benefits, they do not have the same stake in OPEC's future performance unless their dependence upon oil exports should increase drastically. Some of the non-OPEC exporters share the colonial or semi-colonial past of the OPEC countries; others do not. Because the petroleum industry is of less importance to their national economies, the control of that sector is somewhat less relevant to the issue of national independence. Thus, OPEC's success or failure is less a question

---

[1] Jean-Marie Chevalier, *Le nouvel enjeu petrolier* (Paris: Calmann-Levy, 1973), p. 80.

[2] Øystein Noreng, *Oil Politics in the 1980s* (New York: McGraw-Hill, 1978), p. 61.

of their own success or failure. Their future economic development, terms of trade, and standing internationally are much less dependent upon OPEC's performance. Consequently, the non-OPEC exporters are not bound by the same solidarity, and they are unlikely to develop the same ideological cohesion in relation to other petroleum exporters.

However, as noted earlier, the non-OPEC exporters also have benefited greatly from OPEC's success. In all cases there has been an economic benefit in terms of increased oil revenues, constituting a windfall profit for the non-OPEC exporters and to some extent making them political rentiers off OPEC's action. In some cases there have been as well significant political benefits as OPEC's success has permitted changes in oil policy. For example, in the cases of Great Britain and Norway, the price rise and the nationalization of oil in major OPEC countries have improved their bargaining relationship with the international oil industry, permitting a higher level of taxation and a more resolute policy of state participation. In the case of Norway, higher oil revenues also have permitted a more explicitly conservationist depletion policy.

For the non-OPEC exporters, the comparative advantage residing in the fact of being a net exporter of petroleum has been improved by OPEC's success. This gives them a certain stake in OPEC's future performance because the profitability of their petroleum industry and their bargaining position with the international oil industry as well as with the oil importers will depend upon OPEC's performance for some time to come.

Salient Cases: Mexico and Norway

In the 1970s there has been an extensive effort of exploration for petroleum in the OECD countries as well as in the developing nations outside OPEC. So far, two countries appear to have a particular potential for increasing production and exports of oil and natural gas: Mexico and Norway. Both states have expanding production and exports of oil and natural gas as well as a promising potential for expanding known reserves. Both countries have an advantage in that their resources are located fairly close to markets; moreover, they have close economic ties with the countries which are the natural markets for their petroleum. This implies a high degree of security of supply, giving petroleum from Mexico and Norway a political value in addition to its commercial value. Consequently, there is a political rent attached to Mexican and Norwegian petroleum, giving the two governments an especially good position in bargaining with consumers. This fact also makes Mexican and Norwegian policy of particular interest to OPEC.

The oil resources of Mexico and Norway have been subject to much speculation in the international press, and both countries have had the doubtful honor of being pointed out as potential "petroleum saviors" for the Western world. However, at least in a medium-term perspective, the outlook for production is more modest. Mexico is now thought to possess the sixth largest proven petroleum reserves in the world, 40.2 billion barrels of oil and gas.[3] Probable reserves are estimated at 44.6 billion barrels and potential reserves at 200 million barrels.

In Norway, the pace of exploration has so far been slow, confined to a few areas in the southern sector of the continental shelf, and figures should therefore be considered as quite conservative. Present proven reserves in the southern sector are estimated at about 11 billion barrels of petroleum, and the total (proven and probable) reserves are put at about 30 billion barrels. For the northern sector no realistic estimates are available, but this part of the continental shelf makes up about 85 percent of the total, and if the frequency of petroleum here is anything like that in the southern sector, the reserves could be large indeed.[4] It should be added that petroleum production in the northern part of the Norwegian continental shelf is likely to be technologically complex and costly due to depths and weather conditions. In any case, the geology of the area gives reason for optimism.[5] Excluding the northern part of the Norwegian continental shelf, where petroleum production would require long lead times, the probable reserves of Mexico, 45 billion barrels, and of Norway, 40 billion barrels, compare well with those of several OPEC countries. Thus, both Mexico and Norway are potentially fairly significant producers and exporters.

Mexico's oil production in 1978 was 1.4 million barrels/day (b/d), and it is expected to reach 3 million b/d by 1988. Mexican gas production reached 2.5 billion cubic feet/day (cf/d) in 1978; it is estimated to be 6. 8 billion cf/d by 1988. The domestic consumption of oil in Mexico, which in 1978 was about 1 million b/d, is expected to reach 2 million b/d by 1988. This gives an export capacity of perhaps 1 million b/d by the latter year; gas available for exports could be between 0.4 and 2 billion cf/d by 1988. These figures imply that the present, moderate program of oil development be retained.

Norway's oil production in 1978 was about 0.34 million b/d forecast at 0.7 million b/d by 1981, and with the current production program, it might

---

[3]United States, Senate, Committee on Energy and Natural Resources, *Mexico: The Promise and Problems of Petroleum* (Washington, D.C.: Government Printing Office, 1979), p. 17.

[4]A modest estimate for the reserves of the northern sector would be 30 billion (proven and probable).

[5]Norway, Ministry of Industry, *Operations on the Norwegian Continental Shelf, etc.*, Parliamentary report no. 30 (1973-1974), (Oslo: Ministry of Industry, 1974), p. 22.

reach 1 million b/d by the late 1980s. In addition, Norway had gas production of about 1.34 billion cf/d in 1978, which could reach 2.8 billion cf/d by 1981, under the present production program. Norway's domestic oil consumption was on the order of 0.2 million b/d in 1978; it is not expected to increase significantly due to Norway's largely hydroelectric economy. The country has currently no domestic consumption of natural gas. The net exports could thus be about 0.5 million b/d of oil and 2.8 billion cf/d of gas in the 1980s unless substantial new discoveries are made and put into production. For both Mexico and Norway, exports of petroleum already increase total export earnings and improve freedom of action in economic policy; the contribution of petroleum to the two national economies will increase throughout the coming decade. However, for both countries the petroleum sector is not likely to become the dominant factor in the national economy, and in both cases it is a deliberate policy to prevent the petroleum sector from assuming an excessive role.

From the viewpoint of OECD consumers, Mexican and Norwegian petroleum is especially attractive because of the security of supply. This not only gives Mexico and Norway a good bargaining position (because of the political rent noted earlier), it also makes these two nations fairly likely subjects to pressure from major trade partners and allies to increase the level of production. Thus, on the issue of production policy there is a conflict of interest between OECD consumers on one hand and Mexico and Norway on the other. This conflict of interest could become more acute as functions of supply problems in the Middle East and of a more conservationist depletion policy in the OPEC countries.

To Produce or Not to Produce?

To produce or not to produce — that is the question for the oil-exporting countries, inside or outside OPEC. The issue can be seen from three perspectives: the economic; the political; and that of managing the long-term global energy balance. From an economic point of view there is a trade-off between lifting oil in order to invest the income thereby generated and keeping oil in the ground. The options are essentially: (a) to produce oil and invest the resulting capital in domestic economic development; (b) to produce oil and invest in foreign assets; and (c) not to produce oil and invest in oil-in-the-ground.[6]

---

[6]Anwar Jabarti, "The Oil Crisis: A Producer's Dilemma," in *U.S. and World Energy Resources: Prospects and Priorities*, ed. Ragaei El Mallakh and Carl McGuire (Boulder, Colorado: International Research Center for Energy and Economic Development, 1977), pp. 130-31.

The trade-off depends upon the absorptive capacity of the domestic economy and the marginal return on domestic investment, the return on foreign investment, and the expected evolution of the price of oil over the period considered. For Mexico, the high level of unemployment and the high rate of demographic growth should indicate a high absorptive capacity in the domestic economy, but this may depend upon profound structural changes in the Mexican economy. For Norway, the high level of economic development attained before petroleum production began indicates a low absorptive capacity in the domestic economy.[7] The economic recession in Western Europe since 1974, however, has hit other exports and consequently increased the domestic absorptive capacity. Both Mexico and Norway are historical importers of capital and have little experience in handling large-scale investment abroad. In addition, the return upon foreign investment, whether direct or financial, has been low in the post-1974 period; in many cases the return has been negative when corrected for inflation. Thus, for both Mexico and Norway it can appear as economically rational to prefer to invest in oil-in-the-ground beyond a certain level of production (to the extent that the price of oil is expected to increase in real terms), especially if the desire is to avoid abrupt structural changes in the economy.

From a political viewpoint, the problem is to secure the amount and the kind of economic growth that is required to maintain social and political stability without dislocations caused by rapidly rising incomes which can lead to undesirable structural changes. This again has to do with the absorptive capacity of the domestic economy as well as with the existing distribution of income. For Mexico, the unequal distribution of income means that swiftly increasing production of petroleum could lead to rising expectations in the entire population; yet the oil wealth in reality could benefit a minority so that unequal distribution of income, combined with increasingly acute structural imbalances in the economy, could in turn lead to social unrest and political instability.[8] The example of Iran in this respect provides an eloquent warning. For Norway, the high level of economic development and the fairly equal distribution of income mean that quickly increasing production of petroleum could lead to strong inflationary pressures, structural imbalances in the economy, and to a more unequal distribution of income. This, then, could compromise the social stability of the country and produce a political backlash against the rapidly developing oil industry. Consequently, both Mexican and Norwegian governments

---

[7]Norway, Ministry of Finance, *Petroleum Industry in Norwegian Society*, Parliamentary report no. 25 (1973-1974), (Oslo: Ministry of Finance, 1974), pp. 6 ff.

[8]United States, Senate, Committee on Energy and Natural Resources, op. cit., p. 49.

have good political reasons for keeping to a moderate level of petroleum output.

As for managing the long-term global energy balance, the problem is to contribute to making long-term demand and supply trends match and to contribute to a smooth transition from conventional oil into new sources of energy. Again the trade-off between present and future production is concerned. This trade-off partly depends upon the short- and long-term outlook for the balance between energy demand and supply. From 1974 until the Iranian upheavals during the winter of 1978-79, there had been some surplus of oil in the world market, leading to a decline in the real price of oil between 1974 and 1978. However, most serious analyses point out that the long-term trends for energy demand and supply do not seem to match very well on a global scale, and that a serious energy scarcity seems to mounting.[9]

During the latter part of this century and perhaps even during the early part of the next, energy could be a serious bottleneck for world economic development. The looming scarcity particularly seems to affect petroleum and especially conventional oil. The dual need is to encourage conservation of energy and the development of new energy sources, and thus avoid a situation where world oil production peaks without sufficient alternatives having been fashioned. In this perspective, it is a responsible policy for new oil exporters, such as Mexico and Norway, to have an explicit preference for future oil production and not to deplete their oil reserves within a relatively short period of time. The point is that petroleum which is not produced and consumed now is held in the ground for future production and consumption. In this way, the relatively conservationist depletion policies of Mexico and Norway imply that oil reserves are kept for their consumers in North America and Western Europe for a longer period of time than would have been the case with a more rapid depletion policy.

In addition, a more short-sighted depletion policy in Mexico and Norway in the 1970s and 1980s could contribute to depressing the market, with perhaps a negative effect on the price, which in turn could retard energy conservation and the development of new sources of energy as being less economical. Moreover, this would mean rapid pumping of oil from new producers during a period when petroleum appears to be readily available from the traditional oil producers—the OPEC countries—instead of deferring the bulk of the output from new producers until the point in time when production from the traditional exporters is expected to peak. Hence conservation of output policies of Mexico and Norway could lessen the impending energy crunch. Finally, these global energy considerations

---

[9]Workshop on Alternative Energy Strategies, *Energy Global Prospects 1985-2000* (New York: McGraw-Hill, 1977), p. 126.

appear to mesh well with the domestic economic and political considerations of Mexico and Norway concerning petroleum policy.

The question is to what extent these national preferences of Mexico and Norway will be understood and respected internationally. As already noted, by having close economic ties with important petroleum-consuming countries, Mexico and Norway are exposed to potential political pressure to modify their energy policies in the sense of a higher level of output. This is perhaps particularly the case for Norway, with a small population and as a member of the North Atlantic Treaty Organization (NATO) and the OECD, as well as an associate of the latter grouping's International Energy Agency (IEA). Several of Norway's major trading partners are among the world's largest consumers of petroleum, which have sizable and increasing energy needs. Therefore these countries could have an interest in a higher level of production than what is considered to be in Norway's own interest.[10] Some nations could in a more critical position try to exert pressure for a higher output through trade sanctions.[11] By remaining outside OPEC, the non-OPEC exporters, and particularly Mexico and Norway, may perhaps expose themselves to greater foreign pressure by falling outside the framework of OPEC solidarity. This, however, should also be assessed in a dynamic perspective.

To the extent that present Mexican and Norwegian petroleum policies are respected and maintained, the two countries are likely to keep rather diversified economic interests and probably make OPEC membership of little relevance. However, should Mexico's and Norway's present energy policies not be respected and even changed as a result of foreign pressure, the two nations are likely to become increasingly dependent economically upon the oil sector, moving them closer to the situation of the OPEC states, and consequently make OPEC membership more relevant. In any case, even with the present policies, there exist not insignificant common interests between Mexico, Norway, and several other non-OPEC petroleum exporters and the OPEC countries.

Interdependence and Interest Relationships

The non-OPEC exporters and the OPEC countries operate as sellers in the same international market. Experience has taught that this is a rather special market, characterized by very low price elasticities in both the demand and the supply side (at least in a medium-term perspective), so that

---

[10]Norway, Ministry of Finance, op. cit., Appendix, p. 93.

[11]Ibid., pp. 93 ff.

the price of oil is highly sensitive to fairly small changes in supplies or to fluctuations in demand that are not matched by corresponding adjustments in supply. Consequently, the non-OPEC exporters and the OPEC states influence the conditions in which they all sell their petroleum; this reciprocity creates an interdependence between them. Among the common interests are those concerning oil price development, production and depletion policy, and even industrial organization. The reciprocity is so far unbalanced simply because the world oil market essentially is subject to the pricing and production policies of the OPEC members. But the marginal influence of the non-OPEC exporters should not be underestimated, particularly if seen in a dynamic perspective.

The non-OPEC exporters historically have benefited from OPEC's price increases and the nationalization of oil in most OPEC countries. In the future, the petroleum policies of the non-OPEC exporters will be increasingly relevant to OPEC to the extent that non-OPEC supplies will increase their share in the world market. In this way the non-OPEC exporters contribute in determining the market conditions in which the OPEC nations operate. The output policies of the non-OPEC exporters increasingly will be a factor determining the residual demand for OPEC oil in the international market, and consequently the context to which the price and production policies of the OPEC countries will have to refer.

The industrial organization of the petroleum sector of the non-OPEC exporters will determine the supply conditions and the profits of the international oil industry for some of its petroleum and thereby influence its bargaining position with the OPEC countries. Traditionally the oil policies of the more important non-OPEC exporters seem to have been fairly compatible with OPEC's own interest. Mexico was the first country after the Soviet Union to nationalize the oil industry. In the 1960s and early 1970s Norway had a tougher policy in relation to the international oil industry in matters of restrictive licensing, state participation, and taxation than most, if not all, OPEC countries.[12] As early as 1969, in the second round of licensing Norway had a more restrictive position than practically all OPEC states with the possible exception of Iraq. It was only the "oil revolution" of the early 1970s which made the OPEC members appear as champions of radical and nationalist resource management; they still have no monopoly in this respect.

Both Mexico and Norway currently have depletion policies which are more conservationist than those of many OPEC countries, and Norway's restrictive licensing policy can even be seen as a sign of a desire *not* to sit on large proven reserves. Canada likewise has a rigorous conservationist policy

---

[12] Adrian Hamilton, *North Sea Impact* (London: International Institute for Economic Research, 1978), pp. 35 ff.

for natural gas. Against this background it is erroneous to hold that OPEC countries are the only source of "oil nationalism." Indeed, conservationist depletion policies and a high degree of state participation plus a high level of taxation—if not full nationalization—increasingly appear as rational policies for petroleum producers, wherever they are in the world, whatever their level of economic development, and regardless of international economic and political affiliations. This means that OPEC's "oil revolution" to a large extent has been backed up by the non-OPEC exporters.

The lesson is that the non-OPEC exporters and the OPEC countries have basically the same considerations in relation to their petroleum policy. The question of whether to produce or not to produce is also a question of the time horizon desired for the maximization of oil-generated revenues. To a large extent this involves the petroleum resources in relation to population and income needs. Countries with sizable populations and correspondingly large income requirements but relatively small petroleum reserves have a rational desire to maximize revenues over a fairly short period of time, i.e., to pump rapidly at high prices. Nations in the opposite situation with small populations and correspondingly limited income requirements yet relatively large oil and/or gas reserves have a logical drive to maximize petroleum revenues over a longer period of time, i.e., to produce more slowly at perhaps more modest prices. Such a dichotomy fits the OPEC countries relatively well and can explain some of the behavior on the price issue.[13] With the non-OPEC exporters the classification is less simple because of more differentiated economic interests; but the differentiated economic interests also indicate other substantial sources of export earnings and less dependence upon petroleum-generated revenues, which generally work in favor of a more conservationist depletion policy.

In the future, there may be a potential for diverging interests between the OPEC countries and the non-OPEC exporters in matters of the depletion policy of the latter. Whereas the conservationist policies of Mexico and Norway have been of assistance to OPEC in the period from 1974 to 1978, the opposite might be the case in a future situation with increasing scarcity of oil. In such conditions, the unwillingness of non-OPEC exporters with a fairly large resource base, for example, Mexico and Norway, to increase output in order to meet rising international demand for oil could eventually be seen as contributing directly to greater pressures upon the OPEC countries to hike their level of oil production and consequently to pump more quickly than they would do by their own preference. This could be interpreted as a conscious lack of solidarity, enabling nations with more

---

[13]Edith Penrose, "Choices for Oil-Exporting Countries," in *Energy Options and Conservation*, ed. Ragaei El Mallakh and Dorothea H. El Mallakh (Boulder, Colorado: International Research Center for Energy and Economic Development, 1978), pp. 43-57.

diversified economies to extend the life-span of their petroleum while contributing to force countries with less-diversified economies to deplete their major asset fairly quickly.[14] This discussion shows that the interdependence between the non-OPEC and the OPEC exporters might well expand in the future as their common concerns grow more acute. Thus, the common interest in managing the long-term global energy balance and in securing a smooth transition from conventional oil to new sources of energy could become more evident.

An additional set of issues of common concern is more directly related to the petroleum industry. Here an important point is how to finance further exploration and development, including secondary and tertiary recovery. Another relevant issue involves the organization and structure of the international oil market and trade, especially the role of national oil companies in relation to the private multinational oil companies. A third factor has to do with natural gas. Several of the non-OPEC exporters as well as several OPEC countries have substantial reserves of natural gas, which could extend the lifetime of their petroleum production and provide the consumers with additional energy supplies. It is pressing that some understanding on the potential for international trade with natural gas be reached and how this trade eventually should be organized. Finally, the non-OPEC exporters and the OPEC nations have evident shared interests in matters of education, research, and development related to the petroleum industry and to energy in general.

These common concerns are likely to be increasingly evident as petroleum exports build up in some of the non-OPEC exporters and as oil exports from some of the OPEC countries will approach a peak. Thus, it could be in the interest of both sides to intensify consultations and eventually to create a forum for common discussions. Such a step, however, would not be without some political significance on behalf of the non-OPEC exporters.

Policy Choices

The OPEC success and the "oil revolution" in 1973-1974 triggered a political response in the form of the ECG, the Energy Coordinating Group (later transformed into the IEA), largely through the initiative of the United States. It can reasonably be assumed that this was an initiative with multiple objectives. The creation of the IEA not only served the purpose of securing the oil interests of the consumers, its objective was also to serve

[14]Ali A. Attiga, "The Impact of Energy Transition on the Oil-Exporting Countries," in *Energy Options and Conservation*, pp. 1-10.

foreign policy goals of the United States. Since the "oil revolution" there has been a more or less constant temptation in Western Europe, and to some extent in Japan, to strike a separate deal with the oil producers, primarily Arab oil producers and Iran, in order to secure supplies. This could be seen as contrary to the interests of the United States and possibly to the interests of large United States-based multinational companies as well.[15]

The IEA could be seen by some as an attempt to corral other Western oil importers within an institutional framework to secure continued dominance by the United States. Thus, the structure and organization of the international oil market not only serve purposes of rationality and efficiency, they can be instruments of political control.[16] At the outset, the ECG, and later the IEA, were widely—and probably correctly—viewed as an attempt to create a United States-dominated countercartel to OPEC, in order to secure consumer economic interests and United States foreign policy goals and signifying an unwillingness to accept the "oil revolution" as irreversible. More recent experience shows its description should be to coordinate the energy policies of the member countries and to stimulate improved efficiency in energy use. For example, the IEA increasingly appears as a forum where other Western oil consumers attempt to discipline the United States in matters of energy policy, and where the United States energy situation, through extravagant patterns of energy consumption and the withholding of domestic energy supplies, more and more is identified as the world's number one energy problem.[17]

Thus, the IEA now reasonably appears to function in a way quite different from the possible intentions of its creator, being more complementary to OPEC than contradictory, and acting more critically in relation to the United States than as an instrument of United States foreign policy. However, in spite of this modification of the IEA's functioning, the IEA - OPEC relationship still gives an impression of political polarization in the world oil market, with a potential for political conflict in relation to oil supplies and prices. This has created some delicate policy choices for the non-OPEC exporters, particularly for those with close OECD ties and with a close relationship with the United States.

The United States initiative in 1974 to create a counterpart to OPEC was

---

15 Øystein Noreng, op. cit., p. 24.

16 Martin Saeter, "Oljen og de politiske samarbeidsformer," *Internasjonal Politik*, no. 2B, 1975, pp. 397-421.

17 See, for example, Organisation for Economic Co-operation and Development (OECD), *World Energy Outlook* (Paris: OECD, 1977), pp. 18 ff.

evidently of considerable political embarrassment to the OECD countries that were potential petroleum exporters: Canada, Great Britain, and Norway. Canada, Great Britain, and Norway had just benefited greatly from the "oil revolution" in terms of a more profitable domestic petroleum production, higher revenues, and an improved bargaining position. For them to join a consumers' cartel under such circumstances could be seen as a sign of political schizophrenia, particularly for Canada and Norway.[18]

In the spring of 1974 the Norwegian government presented important new documents on its petroleum policy, emphasizing conservation of energy resources, a high degree of state participation, and full national control. This was a policy that in its principles and basic ideology was in line with the policy of the OPEC members. Thus, it is not surprising that the Norwegian government also stressed the community of interests in matters of oil production and prices with the OPEC countries.[19] Thus, joining the IEA without reservations easily could imply compromising the petroleum policy just defined.

IEA aims were explicitly to promote self-sufficiency of energy for the entire IEA area (not for individual countries) and to improve conditions for private investment in energy production in the IEA area. These goals, hardly compatible with the new principles of petroleum policy in Norway, and even in Canada and Great Britain, could be seen as a means of putting pressure upon these same countries to accept higher levels of oil output and a more active presence of private foreign oil companies. After some hesitation, Great Britain joined the IEA without explicit reservations in the autumn of 1974, perhaps on the assumption that IEA's impact upon British petroleum policy would be marginal, and that as a major country Great Britain could easily withstand foreign pressure.

Canada also joined the IEA without explicit reservations, but later made some reservations concerning the long-term program encouraging the member countries' investment in each other's energy resources. However, it was the federal government that joined the IEA, and according to the Canadian constitution sovereignty over energy and natural resources is vested with the provinces. Thus, in practice, the IEA could not get a significant influence over Canadian energy policy.

For Norway, the dilemma was more acute, and it has yet to be finally resolved. The question of IEA membership proved to be very divisive within

[18]Canada profited from the OPEC price rises in the autumn of 1973 by immediately raising the price of oil and gas exported to the United States.

[19]Norway, Ministry of Finance, op. cit., Appendix, p. 87. Moreover, the new British Labor government which came to power early in 1974 immediately started to undertake a revision of that country's oil policy, opting for state participation and a much higher level of oil taxation, adopting principles of policy similar to those of Norway.

the Norwegian administration and within the ruling Labor party. The IEA matter arrived a few years after the also divisive Common Market issue, where the government's attempt to make the country a member of the EEC was defeated in a referendum. This had some relevance for IEA membership, as had general considerations of foreign policy. One argument used in favor of full IEA membership was that otherwise Norway could be seen as slipping out of the system of Western alliances; belonging neither to the EEC or IEA, Norway could be viewed as moving in the direction of gradual neutralization, compromising the nation's security interests. An argument used against full IEA membership was that if such a move was seen by the public as compromising full national control over the petroleum resources and on which there is a wide consensus in Norway, an attempt to force through a full membership of the IEA might eventually trigger a profound political polarization over foreign policy, in turn compromising Norway's role in the Western alliance system and possibly leading to a higher degree of neutralization. The United States embassy in Oslo was particularly active in lobbying in favor of full IEA membership for Norway. The solution chosen was of Norwegian associate membership in the IEA, i.e., participating in all essential activities with full access to information, but without being precommitted to crisis management and keeping full sovereignty over matters of energy policy in general and of petroleum policy in particular.

For Mexico, as a developing country and outside the OECD, the IEA was irrelevant. Instead, the historical experience with oil and Mexico's position in the Third World could make closer links with OPEC quite natural. There is some reason to believe that OPEC membership has been considered by the Mexican government, but that it was rejected because it would compromise relations with the United States and have an adverse effect upon trade with the United States (OPEC membership would imply losing the most-favored-nation status). Thus, economic disadvantages would outweigh the advantages of OPEC membership for Mexico.[20] Given an evident ideological community of thought between OPEC and Mexico, this issue perhaps should not be seen as finally settled.

Recent experience shows that some of the political dilemmas related to oil and foreign policy persist. Great Britain is known to guard jealously her sovereignty in matters of energy policy, both within the IEA and within the EEC. In the IEA Great Britain is now one of the more vocal critics of United States energy policy and its shortcomings; and this possibly has been of some importance in preventing the IEA from becoming a counterpart to OPEC under United States direction. Great Britain also faces an uncomfortable dilemma on production policy. With the present policy,

---

[20]United States, Senate, Committee on Energy and Natural Resources, op. cit., p. 66.

Great Britain is likely to become a net exporter of oil on a modest scale for some time after 1980, but unless major new discoveries are made and brought into production, Great Britain could easily become a net importer of oil again by 1990. This can be seen as a double loss, first by exporting oil at fairly moderate prices in the early 1980s, then by importing oil at perhaps much higher prices in the 1990s. The alternative is to impose production controls once self-sufficiency is attained, thereby lengthening the period during which Great Britain is self-sufficient in oil. However, this could be seen as directly contrary to IEA basic principles and as provocative to other Western oil consumers, not the least the United States which is a fairly important consumer of light North Sea crude.

The recent Norwegian behavior illustrates a persistent dilemma. To a certain extent Norway has tried to balance her IEA association with some OPEC links. Contacts between Norway and OPEC, and individual OPEC countries, are increasing and are more regular and structured. There is so far no direct consultation, but there is a widening exchange of views. In September 1978 the Norwegian government and OAPEC (Organization of the Arab Petroleum Exporting Countries) organized a joint conference in Oslo. Politically, the conference was of considerable significance as both sides stressed the community of interests along with some fairly important exchanges of opinions. For example, from the OAPEC side it was pointed out that the relatively high profit margins for private producers in the Norwegian part of the North Sea permit them to underbid producers of OPEC crude, whose profit margins are less, and thus to stimulate unwarranted downward price adjustments.[21] From the Norwegian side it was pointed out that a future price shock for oil could be detrimental to the interests of oil consumers and producers alike, and that in order to avoid this, a gradual increase in the real price of oil should commence as soon as possible.[22] The real price decline of oil since 1974, and especially since the summer of 1977, was to be regretted and considered irresponsible in the view of Norway; it compromised the economics of Norway's own North Sea petroleum production, as well as having a negative effect upon energy conservation and the development of new sources of energy, thus preparing the ground for a future supply and price crunch. It was widely reported in the Norwegian press that there was a substantial understanding between the Norwegian government and the visiting Kuwaiti oil minister on the question of oil prices. The Norwegian oil minister was reported to have

---

[21] Norway-OAPEC Conference, September 1978, Oslo, comments by Fadhil Al-Chalabi, Deputy Secretary General of OPEC.

[22] Norway-OAPEC Conference, September 1978, address by Bjartmar Gjerde Oil and Energy Minister of Norway.

declared that a 10 percent price increase would help Norway.[23] The Norwegian position was also that higher oil prices were necessary in order to extend the petroleum era and that the decline in real prices was an undue burden on the oil-producing countries.[24]

For Norway this exchange of views was beneficial in drawing forth critical viewpoints from other oil producers. For the OAPEC countries, and through them the OPEC countries, it was perhaps helpful to garner some support on the price issue. At the OPEC meeting in Abu Dhabi in December 1978, when the gradual price increases amounting to 10 percent on the average for 1979 were set, Kuwait is supposed to have played an active role. This does not mean that Norway, or the Oslo meeting, was the instigator of the oil price increase decided at Abu Dhabi, but it does indicate moral support from an outside producer with strong OECD interests for the result arrived at by OPEC. The Norwegian official comment was later that the Abu Dhabi price increase was quite justified. Another result of the Conference was an agreement to hold more regular consultations between Norway and OAPEC as well as OPEC.

In the spring of 1979 Norway was requested by the United States to make a long-term agreement with Israel concerning oil supplies, to compensate for the shortfall of Iranian supplies to Israel. As part of the Camp David Agreement the United States had committed itself to secure Israel's oil supplies, but oil cannot legally be exported from the United States. In order to be able to honor its commitment, the United States government would have to propose new legislation or let another country take over the commitment. Mexico is supposed to have received a similar request, but refused it, despite the fact that Israel was already receiving some quantities of Mexican crude. Given the fairly strong pro-Israeli sentiment in the Norwegian population, and in the Norwegian parliament, the United States might anticipate that the request would be more favorably dealt with by Norway.

However, in the Norwegian government the request to commit Norway to long-term supplies of oil to Israel was met with considerable uneasiness. Reasons included the possible consequences in case of a new war in the Middle East, the fear that the request could compromise Norway's relations with the OAPEC and OPEC countries as well as with the Third World in general, and that such a commitment would make clear to the world and to Norway itself that Norway's position was unequivocally among the Western oil-consuming nations. Norway presently has ambiguous relationships with both the IEA and the OPEC-OAPEC tandem. This can be

---

[23]*Dagbladet*, September 28, 1978.

[24]*Arbeiderbladet*, September 28, 1978.

seen partly as a result of an astute strategy of not getting too closely involved with either side. It also reflects a profound dilemma and split interests in the world oil market, reflecting both "OECD interests" and "OPEC interests." For Norway, both sets of interests are equally legitimate. This permits a special role in the world oil market, that of a mediator between producers and consumers. Such a role appears to be acceptable to OPEC and OAPEC. It is not evident that this is acceptable to all OECD countries, primarily to the United States.

Recent experience has also taught Mexico some of the problems of being a petroleum exporter. From any point of view it is evident that the Mexican energy resources are particularly valuable to the United States, especially as oil supply problems become increasingly likely in other parts of the world. Thus, for Mexico its petroleum policy is closely linked with Mexican policy toward the United States. The oil and gas resources give Mexico a good bargaining position. Still, Mexico has urgent social and economic problems to solve. To a large extent the solution to these problems would be facilitated by the active participation of the United States. As mentioned earlier, the security of supply gives Mexican petroleum a political value in addition to its commercial value. It is quite legitimate for the Mexican government to use this political rent to solve the country's urgent social and economic problems when defining its foreign petroleum policy. It would be illegitimate, from a Mexican point of view, not to do so, as that would imply a waste of political resources. Thus, countries desiring to receive Mexican petroleum should also be expected to contribute to a solution to Mexico's population and employment problems, for example, by accepting immigration and/or industrial exports in order to encourage industrialization.

Recent Mexican experience is that these points have so far not been very well understood in the United States. First, it is a widespread impression in Mexico that the United States wants Mexican oil in large quantities, but no immigration and few industrial exports.[25] Second, even at the level of petroleum exports, the United States has not been accommodating. It is evidently in Mexico's interest to diversify oil exports, whereas natural gas exports have only one obvious outlet, the United States. The negotiations on Mexican gas broke down because of disagreement on the price.[26] But the collapse of the deal can be seen as a sign that the United States is not particularly interested in receiving Mexican gas. United States domestic gas supplies appear more assured than domestic oil supplies, and to import

---

[25]See, for example, George W. Grayson, "Mexico's Opportunity: The Oil Boom," *Foreign Policy*, winter 1977-78, pp. 65-89.

[26]United States, Senate, Committee on Energy and Natural Resources, op. cit., p. 84.

fairly large quantities of Mexican gas could imply a certain effort of adaptation of energy uses. Instead, it appears that the United States has been more interested solely in acquiring large quantities of Mexican oil. Thus, the interests are contradictory.

It is in Mexico's interest not to build the relationship with the United States upon oil alone, rather upon a comprehensive basis which can also assist in resolving its social and economic problems. It seems to have been United States policy not to develop a comprehensive relationship with Mexico, i.e., having to accept immigrants and other imports along with petroleum, but to develop a relationship preferably based upon oil alone. To the extent that oil supply problems appear likely in other parts of the world, the United States can be expected to change its attitude and become more understanding of the Mexican position. Realistically, the United States should not expect to receive large quantities of Mexican petroleum without significantly contributing to the solution of Mexico's social and economic problems. In the meantime, Mexico could become more conscious of her status as a Third World country and increasingly identify itself with OPEC. In a tight oil supply situation, such a move could have a rather persuasive effect upon the United States for the need to comprehend Mexico's problems. There are now obvious signs that the United States is reassessing its entire policy with regard to Mexico. Events in the Middle East are probably not without a certain influence in this case.

To sum up, the non-OPEC exporters having close ties to the OECD area are faced with new and delicate problems of policy. On some points these new problems of policy could rationally imply closer contacts with OPEC, because of the interdependence already mentioned, and because closer ties with OPEC could be seen as an effective instrument of bargaining with OECD countries. The most salient issue is, of course, the price of oil itself. For Mexico and Norway, given the mixture of oil interests and other economic interests, it is in their interest not to have a supply and price crunch, but rather to have a relatively smooth evolution of the price of oil, permitting a gradual transition into new sources of energy within a stable world economy. For Great Britain, given the prospect of self-sufficiency in oil of rather limited duration, it is in that nation's interest either not to have a supply and price crunch at all, or to have it as soon as possible. The point for Great Britain is that the real oil prices in the 1990s should not be higher than in the 1980s, and the sooner energy conservation and the development of new energy sources on a worldwide scale can be effectively stimulated, the better. Another important issue is the production policy, where it increasingly seems to be in the interest of all non-OPEC exporters to extend the lifetime of their petroleum resources. A third concern is industrial organization, where the non-OPEC exporters, including Canada, Great Britain, and Norway, seem to opt for a high degree of state participation.

On all these points there is a certain conflict of interests with the IEA, and in particular with the United States. The IEA, and especially the United States, express the desire for constant energy prices, for a higher production of petroleum outside OPEC, and for a more extended involvement of private capital in the development of energy resources. Unless a major international crisis should occur, this conflict of interest is unlikely to lead to open confrontations. However, the problem is perhaps rather that of a more chronic or creeping oil supply scarcity, and in such a situation it could be argued that incremental supplies from the non-OPEC exporters, such as Mexico and Norway, would be of great help.[27] Under such circumstances, the non-OPEC exporters would naturally find themselves in a politically uncomfortable position. This would require a more explicit trade-off between their oil interests and their other economic interests.

The Iranian crisis during the winter of 1978-1979 has created a pressure situation exhibiting some common features with the supply crunch anticipated in the 1980s by some observers. It seems so far that both OPEC and non-OPEC producers are resisting outside pressures on their oil output policies. Should a more acute crisis arise, appearing less transient than the Iranian crisis of 1978-1979, the pressure upon countries such as Canada, Mexico, and Norway could perhaps reach intolerable proportions. This could easily be counterproductive and trigger a more pronounced "petroleum nationalism." In Mexico, as mentioned, the historical experience with oil and the fact of being a developing country could make OPEC membership a natural option in case of heavy foreign pressure. In Norway, the historical experience of foreign domination and the fact that the country to a large extent is an exporter of raw materials and semi-finished products could make closer OPEC links quite natural.[28] In addition, there is in some segments of the Norwegian Labor party as well as in parts of the administration a considerable ideological sympathy for OPEC. This could be of critical importance in case of strong foreign pressure.

In the meantime, contacts between some of the non-OPEC exporters and OPEC countries are increasing and are becoming more structured. An interesting constellation is building up around the Atlantic Ocean in the form of more regular contacts between the national oil companies of Venezuela, Canada, Great Britain, and Norway. The initiative was at the

---

[27]The Rockefeller Foundation, *International Energy Supply: A Perspective from the Industrial World* (New York: The Rockefeller Foundation, 1978), pp. 5 ff.

[28]Dankwart A. Rustow and John F. Mugno, *OPEC: Success and Prospects* (New York: New York University Press, 1977), p. 115.

outset largely Venezuelan, emanating form Petroven, but now also involves Petrocanada, BNOC, and Statoil. The discussions at this level are politically less delicate than regular consultations on oil policy would be between the governments, and they can be a useful substitute. In recent times there has been as well some Mexican participation through Pemex. This could be a precedent for closer contacts between the national oil companies of the non-OPEC exporters and those of other OPEC countries. A meeting was planned for March 1979, to be hosted by the British Energy Minister Tony Benn, with the oil and energy ministers of Venezuela, Saudi Arabia, and Kuwait along with those of Mexico, Canada, Norway, and Great Britain invited. It would be surprising if the agenda would not have included a discussion of price and production. The meeting was postponed because of the Iranian crisis and an extraordinary OPEC meeting on price surcharges. In any case, the sessions would have meant a breakthrough for contacts between the non-OPEC exporters and OPEC. It only seems a matter of time before such a breakthrough will occur.

The contacts are already contributing to a certain intellectual cross-fertilization between the non-OPEC exporters and the OPEC countries; the significance of the Norwegian-OAPEC conference in this respect has already been pointed out. This not only means that the non-OPEC exporters perhaps will be thinking and acting a bit more like OPEC countries, it also implies that the OPEC nations can learn something from the non-OPEC exporters. In terms of oil policy, marketing, and oil management, there are obviously lessons to be exchanged by both, but on the whole the OPEC states should be able to benefit from the more comprehensive insight in economics and business administration and the greater industrial maturity and expertise of some of the non-OPEC exporters.

Politically, this could have two different effects, depending upon the time horizon considered. In the short run, by acquiring more industrial expertise and more insight into the OECD energy markets the OPEC countries might become more able to assert their immediate interests, in particular to increase the real price of oil and to pursue more successfully a strategy of industrialization based on oil. In the long run, by becoming more closely involved with nations having more differentiated economic interests, the OPEC states might become more empathetic to the complexity of the world economy and its varied requirements. In this perspective, closer links between some of the non-OPEC exporters and the OPEC countries would not only be to the benefit of the two parties involved, but in the long run also to the oil consumers. It could even be argued that it would be in the interest of the consumers to have some non-OPEC exporters join OPEC, perhaps Mexico and Norway, as this could contribute to a broadened OPEC base and outlook on the world's economic and energy problems. This, however, would require a more long-term understanding of the interdependence between OPEC and OECD countries than seems to be prevalent in certain key consumer governments, not the least in the United States.

However, United States policy in relation to OPEC has fluctuated over time. There is some cause to think that until 1973 the United States government encouraged OPEC's action, wanting higher oil prices in order to put an end to the competitive disadvantage created by the lower oil prices in the world market than those in the United States. Only after 1973, when OPEC proved more successful than perhaps anticipated by the United States government, was the organization considered an adversary. This was essentially for political reasons, as the 1973-1974 price rise hit the United States quite moderately. But OPEC's potential impact upon relations with Western Europe and Japan and its precedent for other raw materials exporting developing countries were seen as negative. In the present situation, after the Abu Dhabi price increase, the Iranian crisis, and the spring 1979 surcharges, the United States government is hopefully learning that it is difficult to break OPEC, and that OPEC probably has a fairly smooth road ahead for the rest of this century unless it commits major errors. This could be the occasion for a new direction in United States policy in relation to OPEC. Possibly, within a few years, it could be United States policy to cooperate actively with OPEC, and in this connection it could also be United States policy to encourage important non-OPEC exporters, such as Mexico and Norway, to join OPEC. In such a situation, political obstacles for Mexico and Norway to apply for OPEC membership would fall; their admission to that body would be another question. In any case, given their past record they would be unlikely to accept the role of United States hostages in OPEC.

Finally, under realistic oil supply trends in the world, by the early 1990s several of the non-OPEC exporters could be relatively sizable suppliers, whereas in some OPEC countries oil exports might be in a scaling-down process by then. With present trends and unless major new discoveries are made, oil exports by 1990-1995 would be quite small from nations such as Algeria, Iran, Indonesia, and possibly Venezuela as well. Incremental oil supplies would have to come to a considerable extent from non-OPEC exporters, particularly Mexico and Norway. With these countries remaining outside OPEC there would be a need for close consultations, in order to prevent the oil market from behaving like an ordinary commodity market, that is, with substantial price fluctuations subject to fairly small changes in supply and demand. Experience shows that the price of oil is an important determinant in the stability of the world economy; a highly unstable price of oil would be in no one's interest.

On the other hand, experience also shows that the international price of oil can reasonably be subject to administrative fixing, whether by a cartel of integrated oil companies or by a group of oil-producing countries.[29] Therefore, there is a case for including in OPEC all major oil-exporting nations in order to make it as representative as possible. Furthermore, it is in the interest of the consumers to include in OPEC those exporting states which have differentiated economic interests. Since 1973-1974, estimates for the production and use of other sources of energy, such as coal and nuclear power, have been scaled down considerably. This means that oil is likely to keep most of its economic and political clout at least through this century. In this perspective, the stability of the world economy could benefit from the Organization of Petroleum Exporting Countries being as representative as possible, and therefore perhaps from some of the non-OPEC exporters eventually joining OPEC. In the meantime, the non-OPEC exporters could perhaps form a club, "Friends of OPEC."

---

[29]Fadhil Al-Chalabi, "The Administrable Nature of Pricing OPEC Oil," *OPEC Review*, September 1978, pp. 21-42.

14

THE FUTURE RELATIONSHIP AMONG ENERGY DEMAND, OPEC, AND THE VALUE OF THE DOLLAR

*Jawad Hashim**

Tracing the Background

Coming from a country which is a large petroleum exporter, belonging to a region which plays a significant role in OPEC, and heading an institution highly concerned with monetary issues, for me the subject of this paper is a natural and appropriate choice.

It is time to start building up a line of objective thinking in an area where economic literature has been more than flooded with biases both in academic studies and political comments. An example of the bias is the way a question was put in a press conference after a meeting of the International Monetary Fund (IMF) Interim Committee three years ago. "Should the world accept permanent 30 or 40 billion dollar surpluses in a small group of countries?" The Chairman of the Interim Committee Mr. Dennis Healey responded:

> Well, I remember *years ago* talking to a well-known American television commentator about this problem, and he said that he had just been out to cover a strike of constructors who were building a bridge, I think across the Hudson, and when he asked the chap why he was out, the man replied: "Because I've got five kids and none of them will take a Volkswagen". The fact is that there is an unconscionable waste of this precious resource in some countries, and one of the factors behind the whole problem is the failure of the Western countries to do as much to conserve energy and reduce demand for imports of oil as they might.[1]

*Dr. Jawad Hashim is President of the Arab Monetary Fund, headquartered in Abu Dhabi, United Arab Emirates. The views expressed in this paper are those of the author only and do not necessarily represent policies and opinions of the Arab Monetary Fund.

[1] *IMF Survey*, October 10, 1977.

This answer brings the issue of energy demand to the fore with respect to the allegations constantly made against the OPEC group. But what should the OPEC countries themselves do? Continuing the answer to the same question, the former IMF Managing Director Mr. Witteveen said:

> I think normally there are *two instruments* that are used in order to reduce a surplus: *increasing internal demand and appreciating the currency.* We can certainly say that these oil exporters are increasing their internal demand at an enormous speed, and that development programs are generally very ambitious. Of course, they can't go any further in these countries. And I think appreciation of their currency would not help. It would make the oil more expensive still, of course, for the oil-consuming countries, and it would make the products of these countries still more expensive, so that it would frustrate their plans to diversify their economy.[2]

The so-called "energy crisis" first occurred in the United States. During the period 1967-1969 it took an international dimension, essentially as a result of the worries expressed by other industrialized countries. Such worries may have resulted from the fact that these countries were capable of achieving comparable living levels with practically half the rates of American per capita energy consumption (table 1).

Judging by the energy consumption data for 1972, the year preceding the 1973 dislocations, we find (table 2) that what an American consumes in a year is consumed in the remaining large industrial nations in 2.5 years, in other developed countries in 7 years, by a citizen of the OPEC in 26 years, of Latin America in 13 years, of Asia in 51 years, and of Africa in 95 years.

During the same period other events have taken place in the international monetary system. These events were: the termination on 15 August 1971 of the dollar convertibility into gold, the devaluations of the dollar, and the final breakdown of the Bretton Woods parity system in 1973. We may recall that the first adjustment in the price of oil was that of the Tehran Agreement in 1971. The 1973 action was severely felt because it has been delayed for so long. It was quickly absorbed by the ongoing inflationary era of 1974-1975. The persistence of inflation in the industrial countries had a dual adverse effect on oil-exporting countries: it diminished real oil prices, and it reduced the real value of their surpluses. It should be remembered that such surpluses are essentially due to the response of oil-exporting nations to world needs, rather than to their immediate and future growth requirements.

---

[2]Ibid.

Table 1

PER CAPITA GROSS DOMESTIC PRODUCT (GDP)
AND ENERGY CONSUMPTION FOR SELECTED COUNTRIES, 1974

Country	Per Capita Energy Consumption (kilograms of coal equivalent)	Per Capita GDP (U.S. dollars)	Consumption/ GDP (U.S. = 100)
United States	11,485	6,633	100
Canada	9,816	6,636	85
Belgium	6,637	5,466	70
Sweden	5,804	6,886	49
Federal Republic of Germany	5,689	6,216	53
United Kingdom	5,464	3,407	93
Iceland	5,140	6,329	47
Norway	4,925	5,833	49
France	4,330	5,054	49
Japan	4,839	4,133	68
Switzerland	3,608	7,355	28
New Zealand	3,444	4,417	45
Spain	2,063	2,428	49
Greece	2,048	2,140	55
Argentina	1,861	1,988	54
Chile	1,361	741	106
Mexico	1,269	1,119	88
Panama	846	1,133	43
Peru	650	496	76
India	201	137	85
Ethiopia	31	98	18

Sources: United States Department of Commerce, *The Statistical Abstract of the United States, 1976* (Washington, D.C.: Government Printing Office, 1976), pp. 881-82 and United Nations, *Statistical Year Book, 1976* (New York: United Nations, 1977), pp. 372-75 and 686-88.

Table 2

PER CAPITA ENERGY CONSUMPTION BY REGION, 1972

Region	Per Capita Energy Consumption (kilograms of coal equivalent)	Index (U.S./ Region)
A. United States	11,600	1.0
Remaining Group of Ten and Switzerland	4,598	2.5
Other developed[a]	1,620	7.2
Total developed	5,975	1.9
B. OPEC: Arab members	1,148	10.1
Other members	333	34.8
Total	449	25.8
Other Arab countries	302	38.4
Total Arab countries	553	21.0
Other Latin America	885	13.1
Other Asia	228	50.9
Other Africa	122	95.0
Total developing	362	32.1
C. Grand total	2,144	5.4

Source: The World Bank, *World Tables, 1976* (Washington, D.C.: The World Bank or International Bank for Reconstruction and Development, 1976), pp. 480-87.

[a]Includes Australia, New Zealand, South Africa, and "other Europe."

At the same time, OPEC was accused of being a cartel that is exploiting the rest of the world.

> But one should not overlook intra-OPEC price competition. And if one applies the definition of export cartels adopted by developed countries, OPEC does not pass the test. An export cartel not only must include fixed agreements on prices, but also related agreements in such key areas as production control and marketing sharing. The export cartel must also be responsible for monitoring the activities of its constituent members with a view to policing violations and penalize violators. The OPEC member governments do not perform any of these cartel functions.[3]

Inflation: Causes and Blame

Another myth that has been surfacing periodically in the last few years in the West is the allegation that any increase in the price of crude oil is responsible for the spread of worldwide inflation and for causing havoc in the industrial countries. Such allegations circulate in particular just before OPEC semi-annual meetings.

As is known, the behavior of the United States balance of payments has been causing significant worry for some time. On the one hand, a persistent United States deficit provided the means for obtaining international reserve assets. On the other, such assets became more and more dubious as they accumulated. The former aspect did not help very much, since the majority of increases in assets were acquired essentially by a few large-surplus countries.[4] The latter was highlighted by the 1971-1973 events. The issues of international liquidity and confidence eventually exploded and left OPEC members as well as many other countries in a world of international disorder.

A natural outcome was the recession that was experienced by the U.S. economy as well as certain Western European economies that have strong economic linkages with that of the United States. This recession has induced cutbacks in consumers' spending and a worldwide curtailment in industrial production. For example, between 1974 and 1975 the real GNP in the United States decreased while nominal GNP increased. This increment

---

[3]Zuhayr Mikdashi, *The International Politics of Natural Resources* (Ithaca, New York: Cornell University Press, 1976), pp. 78-9.

[4]According to the International Monetary Fund, *Annual Report 1978*, p. 18-20, the combined current account balance of three major OECD countries — Japan, Switzerland, and the Federal Republic of Germany — amounted to about $22.5 billion. The current account balance of the major oil-exporting nations was expected to total $20 billion in 1978.

was due to a cost-push inflation generated by producers — ultimately to satisfy higher profits and wages — by decreasing the supply available in the market. The recession of 1974-1975, accompanied by high and increasing prices (the so-called stagflation), was induced by endogenous rather than exogenous factors, such as an increase in the price of crude oil by OPEC in 1973. In fact, if these were the causes we would have had a classical case of a traditional recession; this would have saved economists from probing into the puzzling phenomenon. What is still contributing to the hardships of the world is the rapid deterioration of the U.S. balance of payments. The United States trade balance showed in 1977 a spectacular deficit of $31 billion or 1.65 percent of the GNP. Apart from a minor deficit in 1972, this reflected the first sizable deficit in the balance of goods and services of $10 billion or 0.5 percent of the GNP, thus leading to a similar situation in the balance on current account ($15 billion or 0.8 percent). Judging from the series for GNP at constant 1975 prices,[5] this is happening in a period where the rate of growth has been falling back to normal from the 1976 rate of 6 percent to 4.9 percent.

U.S. Role in Energy Demand

Since the late forties, the demand for energy has been increasing steadily at rates ranging between 2 to 4 percent annually. The only period of slackening in demand for energy experienced in the United States and Western Europe was during the years of severe recession in 1974 and 1975. The United States plays a very vital role in the functioning of the world's demand and supply for energy; further, it is the single largest oil importer of the OPEC energy supply. The sheer size of the United States GNP, which is about 35 percent of the world's GNP, is a major element in the direction of supply and demand both at present and in the future.

Due to the excessive rate of energy utilization observed before, U.S. consumption worries many energy experts internationally. In fact, United States demand for energy has been cited as the world's "energy problem number one."[6] For example, after 18 months of intensive debating by the

---

[5]GNP series at constant 1975 prices: in 1972, $1,489.4 billion; in 1973, $1,570.6; in 1974, $1,548.8; in 1975, $1,528.8; in 1976, $1,621.1; in 1977, $1,700.7; in 1978, $1,899.1; and in 1979 $2,132.0.

[6]Address of Dr. Øystein Noreng at the Fifth International Conference on Energy in the 1980s: Conflict or Cooperation?, University of Colorado, Boulder, October 16-17, 1978, as reported in the Christian Science Monitor (International edition), October 23, 1978 and the *Financial Times*, October 26, 1978.

Senate of President Carter's energy bill, the United States did not come up with a viable solution to its energy problem. Moreover, the price of much of U.S. oil remains below the world's level. President Carter was faced by a dilemma, to satisfy American voters and Congress or to allay the Western partners' worries about economic conditions in the United States. In fact, the dilemma is much more deeply rooted; it reflects the conflict between short-term and long-term goals. One branch of the U.S. government stresses less reliance on OPEC oil and a decrease in demand in order to reduce the balance-of-payments deficit as a short-term objective. The other branch, including the Department of Energy, encourages stockpiling of OPEC oil, thereby pushing up demand for it, as a longer-term goal.

It is interesting to note that during the same period, the less-developed world or LDCs suffered from a severe shortage of basic foods and an increase in prices which practically halted growth in the Third World, with the margins going essentially to North America. The action, which took the shape of the World Food Conference, soon lost its drive as prices began to fall somewhat, leaving the LDCs struggling, as in the past, for survival in some cases and to stimulate development for all.

OPEC Domination of Oil Reserves

The world oil reserves are estimated to be between 654 and 848 billion barrels as of January 1978. Of this amount the OPEC countries possess 440 billion barrels or 68 percent. The Arab OPEC members account for half the global reserves; Kuwait and Saudi Arabia together possess one-third, with Iran coming next with its 10 percent share. The life expectancy of oil reserves at 1977 rates of consumption is 47 years for Arab oil, 26 for other OPEC, and 20 years only outside OPEC.

However, the picture keeps changing. OPEC known oil reserves, in spite of the continuous flow of exports, more than doubled between 1960 and 1977 (rising from 218 to 440 billion barrels). The same is foreseen for new producers, whose entry often was made possible through price adjustment. The U.S. Congress Subcommittee on Energy and Power has estimated (table 3) an increased share for OPEC from 61 percent in 1980 to 64 percent in 1990.

Besides the demand and supply of oil, we could focus on another fundamental issue which has been gathering force since the sixties, namely, the unprecedented movements of short-term capital from one financial center to another. So far "swap" arrangements between major financial centers could provide some modest solution; but they are only a palliative rather than a remedy. The latter needs more profound analysis, not the superfluous accusation that OPEC surpluses are the cause of trouble.

Table 3

OIL AND NATURAL GAS LIQUIDS SUPPLY AND DEMAND BALANCE, MARKET ECONOMIES
(in millions of barrels per day)

Item	1980	1985	1990
Demand:			
Consumption	54.8	66.9	76.3
Oil storage	1.9	1.9	---
Total demand	56.7	68.8	76.3
Supply:			
OECD, other developed countries	15.6	17.3	19.1
Non-OPEC, developing countries	5.4	7.7	8.3
OPEC supply	34.6	42.8	48.9
Net exports, Soviet bloc	(0.6)	(0)	(0)
Net exports, Mainland China	(0.5)	(1.0)	(0)
Net exports, Socialist Bloc	1.1	1.0	(0)
Total supply	56.7	68.8	76.3
OPEC share in total supply	61%	62%	64%

Source: Library of Congress, Congressional Research Service, *U.S. Energy Demand and Supply, 1976-85, Limited Options, Unlimited Constraints*, Prepared for the Subcommittee on Energy and Power of the Committee on Interstate and Foreign Commerce, United States, House of Representatives (Washington, D.C.: Government Printing Office, November 1977) p. 10.

Before 1973, OPEC had a small surplus in its balance of payments with the industrial countries of the West and Japan. Between 1950 and 1970 OPEC experienced only a small surplus in its balance on current account. The significant rise occurred in 1974, when the surplus amounted to $65 billion (table 4). In the following year it fell sharply to less than $31 billion and to $9.5 billion in 1978. A number of OPEC members, especially those with relatively larger populations (such as Indonesia, Nigeria, Algeria, and Venezuela) are already running deficits on their current accounts.

Since 1973, many have spoken about the enormous increase in the surpluses and reserves held by OPEC, as endangering the essential social and economic fabric of the West. Many econometric studies also projected large foreign exchange assets accumulated by OPEC as possibly choking the international financial community. The main trouble with those models and expert views is that they ignored the role that has been played by multinationals and commercial banks in the Euromarket and their impact on world finance and stability. At the same time, all those studies ignored the real income effect of the surplus, the real value of the dollar since 1973, and the security of OPEC assets invested in the West, especially after the freezing of the Iranian assets by the United States on November 14, 1979. The 1978 dollar is equivalent to 0.7 of the 1974 dollar (table 5) due to the 30 percent drop in its value as measured against the price index of OECD exports. Between September 1977 and October 1978, a further drop of 16 percent occurred, due to the depreciation of the dollar vis-à-vis world leading currencies. During the same period the price of oil remained constant.

Use and Dispersion of Financial Surpluses

The second question to address is: What has OPEC done with its financial surplus since 1974? Here it can be easily seen that OPEC demonstrated responsibility towards the international financial order buy seeking a stable return on their investment rather than a quick turnover with speculation in international currencies. Nor have OPEC nations taken over or dominated Western industries as feared by the Western press and in political and academic circles. For example, Saudi Arabia has stated that it seeks to play a constructive role, recognizing the need to act with larger issues in mind than solely profit. In this regard, it has sought to avoid sudden or large-scale shifts in assets, speculative transactions, investment in the sensitive area of real assets and controlling interests, especially in the United States where they hold the largest single component of their international reserves.

According to the U.S. Treasury, while Middle East oil producers' assets

Table 4

OPEC BALANCE ON CURRENT ACCOUNT
BETWEEN 1971 AND END OF 1980
(in billions of U.S. dollars)

Year	Oil Exports[a]	Nonoil Exports	Imports[b]	Trade Balance	Net Service	Net Official Transfers	Current Account Surplus
1971				9.3			2.3
1972				10.0			2.0
1973	34.9	4.6	20.5	19.0	-12.4	-0.9	5.7
1974	112.9	5.6	36.6	81.9	-14.8	-2.5	64.7
1975	104.4	5.6	56.3	53.7	-18.3	-5.1	30.4
1976	126.4	7.2	67.8	65.8	-24.8	-4.7	36.3
1977	138.5	9.0	82.0	65.5	-30.5	-5.5	29.5
1978	131.0	10.5	95.5	46.0	-31.5	-5.0	9.5
1979	164.0	11.0	128.0	62.0	-34.0	-8.0	55.0
1980	207.0	12.5	155.0	69.0	-38.5	-11.5	95.5

Sources: Organization for Economic Cooperation and Development (OECD), *Economic Outlook* (Paris: OECD, July 1978) and Citibank, October 1978.

[a]Oil exports assumed to increase during the years of 1979-1980 by 14 percent and the prices to increase by 11 percent, respectively.

[b]Years of 1979-1980 are estimated; 1979 imports increased by 30 percent annually and prices by 15 percent while for the year of 1980 it has been estimated to increase by 21 percent annually.

Table 5

PRICE INDICES TO OPEC AND OECD EXPORTS
BETWEEN 1970 AND END OF 1980
(1974 = 100)

Year	1970	1973	1974	1975	1976	1977	1978	1979[a]	1980[b]
OPEC exports ..	17	31	100	107	113	125	126	141	157
OECD exports ..	58	80	100	112	113	125	142	165	189

Sources: International Monetary Fund (IMF), *International Financial Statistics* (Washington, D.C.: IMF, 1977 and 1978) and W.J. Levy, "Recycling Surplus Petrodollars via Internationally Issued Indexed Energy Bonds," *Middle East Economic Survey*, April 7, 1980, pp. 1-7.

[a]1979 OPEC exports prices were assumed to have increased by 11 percent, while imports prices increased by 15 percent.

[b]1980 prices were projected with the same annual percentage rate of 1979.

in the United States are large in an absolute sense (about $36 billion at end 1977), their total holdings are very small in comparison with the $3.3 trillion total size of the U.S. capital market. Although their investments are concentrated in Treasury securities, Middle East investors account for less than 10 percent of all foreign holdings of U.S. Treasury securities and less than 3 percent of total public holdings. Both on a stock basis and on an annual purchase basis, their investment in corporate bonds and equities represent less than 1 percent of the total outstanding value and dollar volume of each of these types of U.S. securities. Similarly, their deposits in all United States banks (including their branches abroad) account for only about 5 percent of the total deposits of the large institutions and only about 2.5 percent of the deposits of all commercial banks. In a sense, the OPEC investments have played an important role in financing the United States deficit so far, thereby contributing to financial stability in a highly volatile world. The OPEC investors have paid the price in the shape of continous depreciation of their assets which to them means depletion of the major wealth on which they heavily rely for their future development.

It is true that the bulk of the the OPEC surplus of 1974 was invested in short-term assets, especially in the dollar market. But in subsequent years, OPEC's investment preferences shifted to long-term fixed-interest securities and long-term bank deposits. In both cases the preference was for dollar-denominated investments. For example, by the end of 1977, OPEC's total investment in the dollar market amounted to $87 billion, of which $36 billion was in the United States itself. By October 1978, the dollar-denominated investments increased by another $7 billion (out of a total $25 billion) in spite of the decline in the value of the dollar during that year. In the last two years, a large portion of OPEC's investment of its external assets has been channelled as loans and grants to the OECD nations, international organizations, and less-developed countries. The total amounted to $42 billion by the end of 1977 and increased to about $45 billion by the end of 1978. By and large, the choice of OPEC investment has been the dollar market because of its potential and volume in comparison with other markets, even with the decline in the purchasing power of the dollar in relation to other Euro-currencies. Since the end of 1979, OAPEC (Organization of the Arab Petroleum Exporting Countries) has been moving in the direction of diversification of their assets after the Iranian experience with the freezing of its assets in the United States and the U.S. commercial bank branches overseas. Thus, rather than being the cause of instability of the dollar in the international market as frequently alleged, the OPEC investments were an important stabilizer.

The share of petro-dollars in the Euromarket is only a small proportion of the total volume, about 13 percent. It is natural to deduce that the 87 percent would influence the 13 percent rather than vice versa. According to

Professor Robert Triffin, the Euromarkets are contributing to the explosive growth of world liquidity. Mr. Michaele Faratiani, Common Market Commission economist, stressed the fact that the Euromarkets are uncontrollable, and they can be used to frustrate domestic economic policy in any country. There has been agreement among many economists that the size of the Euromarkets and the large proportion of the dollar implies that any small change in the portfolio composition of assets will lead to a substantial fluctuation in the value of the dollar. All these facts call for a concerted effort by Western nations, businesses, international organizations, and OPEC to bring order and stability to international finance. As indicated some at the World Bank and IMF, the mere appeal for world economic cooperation is not enough. There should be a call for measures to correct domestic economic policies, to reduce inflation and unemployment, to provide a viable energy policy, and to control deficit spending in the United States. The important thing is to correct overall as well as regional imbalances. The United States has a deficit with OPEC, yet receives a capital inflow from it; but it has a deficit with OECD states and suffers from a net capital outflow to them.

Conclusion

The future demand for energy leaves the world for a number of years yet as dependent on OPEC's oil as it has been in the recent past, in spite of energy conservation measures and an evolving energy system. At the same time, OPEC's future role in the world is not to choke off world economic growth but to protect both its exhaustible resources and the depreciating price of its product in terms of the decreasing purchasing power of the dollar associated with the acceleration of industrial prices in the West. Other future concerns would be in relation to OPEC assets in the West and its value in a world dominated by the depreciating dollar. It is plausible to suggest future world trade be tied to a numeraire which does not depreciate in value through times in order to protect international trade and finance from wild fluctuation.

15

OPEC REVENUES AND INFLATION IN OPEC MEMBER COUNTRIES: A FISCAL POLICY APPROACH

*Massood V. Samii**

The structure of the economy in OPEC member countries differs from that of many other nations due to the heavy dependence on the export of one primary, depleting commodity: petroleum. The oil sector provides not only foreign exchange but also the needed funds for investment. The increase in oil revenues of these countries, as a result of the oil price rise in 1973, provided an engine of growth that enabled many OPEC states to plan a very high rate of economic growth without having to be overly concerned with financial resource constraints. However, nonfinancial constraints began to affect the plan's implementation in these countries. Among the most serious constraints were manpower and infrastructure limitations, which caused considerable cost overrun in project implementation.

A severe rate of inflation is one of the manifestations of the relatively high rate of economic growth in OPEC member countries. While inflation was relatively unknown in these countries prior to 1973, after that year it became one of the major economic problems (see table 1). Despite various measures taken by many countries, including wage-price controls, until 1978 quite a few of them had double-digit inflation, and OPEC as a whole was experiencing price rises of close to 15 percent annually.

It is important to observe that improvement in the standard of living of OPEC member countries (as reflected by the increase in per capita income) implies not only a *quantitative* increase in consumption, but also a *qualitative* improvement; the latter cannot be totally isolated from the general increase in the cost of living – such isolation would require a degree

---

* Massood V. Samii is an Economic Analyst with the Organization of Petroleum Exporting Countries in Vienna. Prior to assuming his present post, he was Director of the Management Division of Western Consultants, Iran, and had served as Acting Director of the International Economic Bureau, Plan and Budget Organization, in Iran. The author holds a Ph.D. in economics from the State University of New York at Albany and has taught at colleges and universities in both the United States and Iran. The views expressed here are those of the author and do not necessarily represent those of OPEC.

Table 1

INFLATION RATES IN OPEC STATES[a]
(percent change in consumer price index)

Country	1970	1971	1972	1973	1974	1975	1976	1977	1978
Algeria	6.7	2.6	3.6	6.3	4.6	8.2	9.5	11.9	17.2
Indonesia	12.3	4.4	6.5	30.9	40.7	19.0	19.8	11.0	8.1
Iran	1.6	4.2	6.5	9.8	14.3	12.7	11.3	27.3	11.6
Iraq	4.4	3.5	5.2	4.9	8.3	9.5	10.3	7.7	n.a.
Kuwait	n.a.[b]	n.a.	n.a.	8.4	13.2	8.9	5.5	8.2	9.0
Libya	-5.2	-3.1	-0.3	8.0	7.4	9.2	5.4	6.4	n.a.
Nigeria	13.8	16.1	2.6	5.7	12.5	33.7	22.0	21.5	24.4
Saudi Arabia	0.2	4.4	4.4	16.6	21.4	34.6	31.6	11.3	-1.6
Venezuela	2.6	3.3	2.8	4.1	8.4	10.1	7.7	7.7	7.2
OPEC	5.2	4.9	4.2	10.9	16.2	17.9	15.4	14.9	8.9

Source: International Monetary Fund, *International Financial Statistics*, various issues.

[a]No data were available for Ecuador, Gabon, Qatar, and the United Arab Emirates.
[b]Not available.

of disaggregation almost impossible to attain. Thus, at least a part of the increase in the consumer price index must be attributed to the qualitative factor.

Inflation: Some Contributory Factors

There are various factors to which inflation can be attributed, including increases in the price of imported goods, physical bottlenecks, shortages of labor, and government fiscal policies.

OPEC member countries claim that one of the major factors contributing to domestic inflation is imported inflation. Prices of imported consumer goods enter into the consumer price index, both directly and indirectly. Since a portion of consumer goods in these countries is imported, the direct effect accords with the share of imported goods in the total consumption basket. The indirect effect is the amount of increase in the price of domestically produced goods as a result of increase in the cost of imported intermediary goods.

OPEC has been monitoring the import price index of member countries for several years and has revealed a steep rate of increase.

The relevant index for goods imported from the industrialized countries shows

> that unit values increased in 1974 alone by 32.6% over 1973. By the end of 1978 this index had reached 324 on the basis of 1973. Naturally there is a wide variation between OPEC Countries.[1]

The reported import index of OPEC is much higher than the export price index of industrialized countries. One reason is that the first index is based on c.i.f. and the second on f.o.b. (that is, inclusion or exclusion of freight charges). Also, the export baskets of industrialized nations have a different commodity mix than the import basket of OPEC states. Moreover, the import price index of OPEC does not take into account improvement in the quality of imported goods.[2]

Another inflationary factor in the economy of OPEC states has been the existence of various physical bottlenecks, such as port congestion, lack of adequate transportation facilities, and outdated distribution channels. These problems have caused sectoral cost increases, which have eventually been reflected in the price to the final consumers.

A further contributing element has been the shortage of labor in these countries. Implementation of ambitious development plans in various OPEC states requires a considerable labor force, especially of skilled workers. Although a few member countries, such as Indonesia and Nigeria, have an abundance of labor, most other OPEC nations are sparsely populated. Even those in the former category lack trained manpower for the implementation of sweeping development projects. The implication of market forces has resulted in an outward shift of the demand curve for labor and, hence, an upward pressure on the wage rate. Imported labor, although eliminating bottlenecks and increasing labor supply, has failed to prevent wage increases. Actually, the higher cost of foreign labor in itself has been a contributing factor to the demand of indigenous workers for higher wages.

Another element causing inflationary pressure on the economy of OPEC countries has been government fiscal policy based on revenues from the oil sector. In fact, this policy could be considered the main contributing factor to inflationary pressures in these countries. In the following section, the effect of oil revenues on fiscal policy and the resultant inflationary pressure will be analyzed and various policy tools discussed.

---

[1] Adnan A. Al-Janabi, "Equilibrium of External Balance Between Oil Producing Countries and Industrialized Countries," *Middle East Economic Survey,* Supplement, November 19, 1979, p. 3.

[2] Ibid.

A Formal Structure for Analysis

In order to analyze the inflationary effect of government fiscal policy, it should be disaggregated into its two components: foreign exchange and domestic currency. Each one separately and then in interaction should be analyzed in order to gain insight into the inflationary impact of OPEC's fiscal policy.

The foreign exchange component of government revenues is basically from oil exports, and government foreign exchange expenditures are for import of consumer and investment goods. Furthermore, governments sell foreign exchange to private sectors for their imports. The foreign exchange need of private sectors is determined by excess imports over exports of these sectors.

The domestic currency component consists of government revenues in the form of national currency – including taxes, revenues from state enterprises, and sales of foreign exchange for import to private sectors – minus government domestic expenditures for purchase of consumption and investment goods.[3]

The import by private sectors is assumed to be a function of exchange rate and the ratio of price of domestic goods to the price of imported goods.

$$GFB = OR - (C_{gimp} + I_{gimp}) - SFP \quad (1)$$
$$GDB = T + SFP + GR - (C_{gd} + I_{gd}) \quad (2)$$
$$SFP = IM_{pr} - E_{pr} \quad (3)$$
$$IM_{pr} = f\,(ER, P_d/P_f) \quad (4)$$

when:

GFB = government foreign balance
OR = oil revenues
C_{gimp} = government import of consumption goods
I_{gimp} = government import of investment goods
SFP = sale of foreign exchange to private sector
GDB = government domestic currency balance
T = tax revenues
C_{gd} = government expenditures domestically on consumption goods
I_{gd} = government expenditures domestically on investment goods
GR = government domestic revenues (excluding taxes)

[3] This breakdown is different from the traditional separation of government expenditures to current and development budget. However, it does cover various categories of government expenditures.

IM_{pr} = import by private sector
E_{pr} = export by private sector (nonoil exports)
P_d = price of domestic goods (domestic price index)
P_f = price of foreign goods (import price index)
ER = exchange rate
λ = coefficient relating government imports to oil revenues
γ = coefficient relating government domestic expenditures to oil revenues.

Equations (1) and (2) demonstrate the government's foreign exchange and domestic component balance, respectively. In the budget, the sum of GFB and GDB demonstrates a surplus or deficit of the government. However, aggregation of GFB and GDB hides, as will be demonstrated subsequently, the inflationary (or deflationary) effect of government fiscal policy.

We assume that government import and domestic expenditures are both functions of oil revenues. The functional relations are shown by λ and γ, respectively.[4] Therefore, equations (1) and (2) reduce to the following:

$$GFB = (1 - \lambda)\ OR - SFP \qquad (5)$$
$$GDB = T + GR + SFP - \gamma OR. \qquad (6)$$

It is further assumed that $\lambda+1$, which indicates the net foreign trade of OPEC governments, is positive.[5] This, however, does not mean that GFB is always positive, because imports by the private sector could increase to a point where they cause a negative overall balance of trade.

Figure 1 demonstrates the relationship between GFB and GDB. The foreign balance has a positive slope $(1-\lambda)$ with respect to oil revenue. This means that if oil revenues increase (assuming that SFP remains constant), the foreign balance of the government improves. It should be noted that λ depicts the sensitivity of government imports with respect to oil revenues.

GDB also has a functional relationship to oil revenues; however, it can be seen from equation (6) that this relation is inverse, i.e., as the oil revenues increase, the domestic balance of the country decreases. The equilibrium point is where both the foreign exchange as well as the domestic currency compo-

---

[4] Testing the relation between government imports (G_{imp}) and government domestic expenditures (G_{dexp}) each with oil revenue, using Iranian data, the following values were obtained:

$$G_{imp} = 210.98 + 0.46\ OR \qquad R^2 = 0.98$$
$$G_{dexp} = 508.95 + 0.84\ OR \qquad R^2 = 0.96,$$

that is, $\lambda = 0.46$ and $\gamma = 0.84$. The intercepts cause a parallel shift and do not affect the analysis.

[5] See M. V. Samii, "Economic Growth and Optimum Rate of Oil Extraction," *OPEC Review*, autumn 1979.

Figure 1

RELATIONSHIP BETWEEN GOVERNMENT FOREIGN BALANCE (GFB) AND GOVERNMENT DOMESTIC CURRENCY BALANCE (GDB)

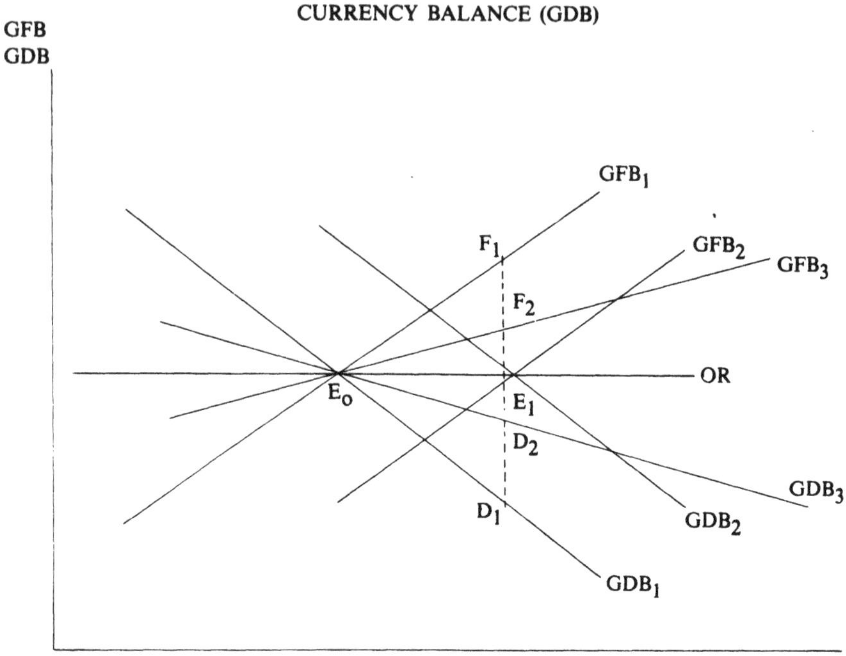

nent of the budget are balanced. Such a point in figure 1 is shown as E_0, where both GFB_1 and GDB_1 are equal to zero.

Suppose the economy is initially at the point E_0; then there is a sudden increase in oil revenues as a result of oil price increases, which means that oil revenues increase from E_0 to E_1. It follows that there will be a positive foreign exchange balance E_1F_1. However, the government, as a result of the increase in its oil revenues, will increase its domestic expenditures by an amount equal to E_1D_1.

It should be noted that although the overall budget of the government at point E_1 might show a small amount of surplus or deficit (equal to $E_1F_1 - E_1D_1$), each of the two balances is at disequilibrium. The overall deficit or surplus in the government budget for any given level of oil revenues depends upon the relative steepness of GDB, (γ), with that of GFB, ($1-\lambda$). It is easy to show that it is not the overall government budget deficit (or surplus) that creates inflationary (or deflationary) pressure, but rather the deficit in the government domestic balance E_1D_1 that is the main cause of inflation.

Government Fiscal Policy and Inflationary Pressure

In order to analyze the inflationary effects of oil revenue increases on the economy of OPEC member countries, the relationship between aggregate supply and demand should be looked at. More specifically, a positive difference between aggregate supply and demand implies inflationary pressure; a negative difference implies deflationary pressure; balance implies neither inflationary nor deflationary pressure. This is shown in equation (7):

$$dP/P = f\ (AggD - AggS) \qquad (7)$$

where the left side of the equation demonstrates price change or inflation as a function of the right side, which is the difference between aggregate demand and aggregate supply. The aggregate demand consists of the sum of the private sector's consumption and investment, the public sector's consumption and investment from domestically produced and imported goods, plus exports minus taxes and other sources of government revenue:

$$AggD = C_p + I_p + C_{gd} + I_{gd} + C_{gimp} + I_{gimp} + E - T - GR. \qquad (8)$$

Aggregate supply, on the other hand, is defined as the sum of domestically produced goods and services and imports of private and government sectors. The aggregate supply can be written as:

$$AggS = y_d + IM_{pr} + C_{gimp} + I_{gimp}. \qquad (9)$$

The inflationary (or deflationary) pressure can be obtained by transposing equation (8) and (9) into equation (7):

$$dP/P = f\ (\lambda OR - T - SFP - GR). \qquad (10)$$

It should be noted that in obtaining equation (10) it was assumed that the private sector's consumption and investment were equal to its output (y_d). This assumption implies that the main function of government is administrative work that does not create output. The value inside the parentheses is the right-hand side of equation (10) and depicts the government domestic balance. Therefore, the above equation confirms the proposition put forward earlier, namely, that the inflationary pressure in the domestic economy is generated by the deficit in the government domestic balance.

Policy Implications

The oil price increases in 1973-1974 and again in 1979 created a relatively large rise in the oil revenues of OPEC states. Moreover, all indications from the international oil markets point to the fact that this trend will continue into the future. The above model provides an insight into the various policy instruments that could be utilized in order to minimize the inflationary impact of expected increases in the oil revenues of OPEC member countries.

Assuming that initially there is an equilibrium at point E_0 in figure 1, an increase in oil revenues in E_1 creates a disequilibrium in both the foreign exchange balance and the domestic currency balance of the government. The disequilibrium will be eliminated as a result of a shift to the right of GFB_1 and GDB_1 to GFB_2 and GDB_2. Assuming that there is no change in government policy, especially with regard to exchange rate revaluations, the adjustment will be the result of relative price changes.

The deficit in GDB_1 at point E_1 transfers itself to an increase in domestic prices, as can be seen from equation (10). Higher domestic prices relative to imports will lead to an increase in imports by the private sector (equation (4)), and hence an increase in demand by the private sector for foreign exchange, SFP. The result will be a shift to the right of both GFB_1 and GDB_1, as shown by equations (5) and (6), which means that although the equilibrium is achieved, this has been at the expense of a high rate of domestic inflation.

However, with the use of available policy instruments, the government could direct the system toward equilibrium without creating much of an inflationary impact. The most effective instrument is the use of exchange rates. The revaluation of domestic currency will increase imports by the private sector, thereby implying a decrease in the government foreign balance and an increase in its domestic balance; i.e., revaluation will cause a shift to the right of both GFB_1 and GDB_1. A correct par value between domestic currency and foreign currencies will result in achieving equilibrium in both balances. However, the use of foreign exchange in OPEC countries might have dynamic adverse effects.

Although in my model I have assumed export to be exogenously determined, in reality it is a function of exchange rate. Therefore, revaluation not only increases imports, but also decreases exports. Of course, this is in line with the objective of obtaining equilibrium in the two balances; however, a decrease in exports contradicts the stated long-run objective of most OPEC states, specifically, an increase in as well as diversification of exports. In order to make exports from these countries more competitive in the international market, a devaluation is necessary. Therefore, an evaluation of short-run versus long-run objectives is essential for OPEC governments in order to determine the optimum level of exchange rate. It should be mentioned that in those countries with an abundance of foreign exchange, a floating exchange rate system would greatly favor the short-run objective.

Another policy that could be implemented by the government is the curtailment of oil production to the level dictated by the absorptive capacity of the economy. The optimum rate of oil production has been discussed by the author elsewhere,[6] and it is not the objective of the present discussion to

[6] Ibid.

enter into that debate. However, the appreciation of the value of a barrel of oil since 1970 has been such that, with the exception of gold, no other asset could be compared to it. A decrease in oil production, as the result of price increases, to a level that would increase oil revenues only marginally would move the system gradually toward equilibrium with relatively low rates of inflation. This process would provide the government with an additional instrument, during a period of oil price stability, to satisfy the ever increasing need of the economy for foreign exchange.

Another possibility is the structural change in the pattern of government expenditures. Although it was assumed that there is a stable relationship between government imports and government expenditures with oil revenues (equations (5) and (6)), a conscious governmental effort toward increasing its imports would reduce the adverse effect of oil revenue rises on the domestic and foreign exchange components of the budget. This policy is demonstrated in figure 1 by rotation of GFB_1 to GFB_3 and GDB_1 to GDB_3. The result is that not only is the domestic balance in this case D_2E_1 smaller than in the reference case E_1D_1, but also the foreign balance E_1F_2 is smaller than E_1F_1.

A word of caution is needed here: There is a limitation regarding the absorptive capacity of the domestic economy with regard to imported goods. A rate of increase in imports that is too fast might lead to inefficiency and waste due to the lack of infrastructure. This was, in fact, experienced by some OPEC countries after the price adjustment of 1973, when port congestion led to huge demurrage charges and, hence, considerable financial losses.

Summary

This chapter has discussed the effect of oil revenue increases on the fiscal policy of OPEC countries. We have seen that the relatively moderate rate of inflation in these countries began to escalate after 1973 to a double-digit level for most OPEC states; for a few countries it exceeded the level of 30 percent annually for a number of years. Various causes of inflation in these countries were discussed, among them imported inflation and physical and manpower bottlenecks. However, it is claimed that the government fiscal policy has been the main factor causing inflationary pressure in these nations.

A frame of analysis, in the form of a model, was considered to determine the effect of oil revenues on the government budget and, in turn, on the price level of OPEC countries. I have suggested that the balance in the government budget conceals, to a great extent, the inflationary pressure resulting from the increase in oil revenues. Since the positive foreign balance and negative domestic balance partially cancel each other, the net balance is naturally very small. I have shown that it is, in fact, the negative domestic balance of the government which creates an inflationary pressure on the economy.

The mechanism for the return to equilibrium has been demonstrated as being domestic inflation. The higher rate of increase in the domestic price level relative to that of imported goods would lead to an increase in imports by the private sector, and hence, a shift in both domestic and foreign balance of government to equilibrium.

However, there are other policy instruments, such as exchange rates, oil output level, and government propensity to import, that could be used to prevent or substantially reduce the inflationary impact of oil price increases. The various effects of using each of these policy instruments have been discussed. None of the instruments could be considered superior. Depending upon the priority of the government with regard to a short-term versus long-term objective, a mix of various instruments should be utilized.

The requirement for the government domestic currency and foreign exchange budget to be balanced implies initially a need for alteration in government budgeting practices toward explicitly separating various types of revenues and expenditures.

16

INFLATION, DOLLAR DEPRECIATION, AND OPEC'S PURCHASING POWER

*Mansoor Dailami**

Since at present the United States dollar serves both as a denominator for pricing crude oil and as a medium of payment between the oil-importing nations and the members of the Organization of Petroleum Exporting Countries (OPEC), then any movement in the value of the dollar against other currencies affects the real export earnings of OPEC. Any downward movement in the value of the dollar in the foreign exchange market results in an erosion of OPEC purchasing power making the OPEC imports originating outside the United States more expensive. Conversely, any upward movement in the value of the dollar increases the implicit price of oil and thereby results in a windfall gain in OPEC's fortune.

Recently much concern has been expressed on the impact of the dollar's depreciation on the real export earnings of OPEC and the implications of any protective action taken by OPEC, either by raising the dollar price of oil or by changing its pricing mechanism to a currency basket, on the future stability of the dollar and on world economic conditions.[1] With approximately 80 percent of OPEC imports originating outside the United States and with a predominantly large proportion of OPEC's past accumulated surpluses invested in dollar denominated assets, the loss

---

*Mansoor Dailami, presently with the Massachusetts Institute of Technology Energy Laboratory, holds degrees in economics from the University of London (M.Sci.) and Harvard University (Ph.D.) with specializations in econometrics, international trade and finance, and Middle Eastern development economics. Among his working papers at the Energy Laboratory are "The Determination and Control of Money Supply in an Oil-Exporting Country: The Iranian Experience, 1961-1975" (MIT-EL-78-027 WP) and "The Choice of an Optimal Currency for Denominating the Price of Oil" (MIT-EL-78-026 WP). The author wishes to thank Dr. James L. Paddock for discussion and helpful comments on this study.

[1]This is calculated on the basis of total dollar holdings of OPEC of about $96 billion at the end of 1977, O. Aburdene, "1 Petrodollar = 72 Cents," *Euromoney*, May 1978.

incurred as the result of dollar depreciation appears to be substantial. For instance, a uniform 10 percent depreciation of the dollar against the currencies of other industrial countries will result in a loss of about 8 percent in OPEC's real purchasing power and in a loss of about $7.6 billion in asset value as calculated on OPEC's total foreign investment at the end of 1977.[2]

The objective of this paper is to provide some empirical analysis of the impact of dollar's fluctuation on OPEC's terms of trade over the period 1971-1977, and to assess to what extent the decline in OPEC's terms of trade, after the fourfold oil price increase of late 1973, can be attributed to the falling value of the dollar and to what extent to the high rates of inflation prevailing in the industrial countries. The study is divided: a theoretical analysis of OPEC's terms of trade (the model), the empirical results, and a brief summary with some significant conclusions.

The Model

Assuming oil is the only export of the OPEC countries, the terms of trade, ti, of the ith member country can be written as

$$ti = Po/Pmi \qquad (1)$$

where Po is the price of oil[3] and Pmi is the unit import price for ith country both expressed in terms of the same currency (dollars in this case).

From (1) the rate of growth of ti can be written as the difference between the rates of growth of Po and Pmi, that is,

$$\dot{t}i/ti = \dot{P}o/Po - \dot{P}mi/\dot{P}mi \qquad (1.a)$$

where (.) on the top of any variable indicates the time derivative of that variable.

To derive the unit import price index for ith country we first write the total value of its imports in United States dollar as,

$$Vi = \sum_j Xij\ Pj\ Rj \qquad (2)$$

where Xij is the quantity of imports of ith OPEC country coming from jth country, Pj is the export price of jth country expressed in terms of its domestic currency, and Rj is the exchange rate of jth country defined as the

[2]See, for example, J. Amuzegar, "OPEC and the Dollar Dilemma," *Foreign Affairs*, July 1978, pp. 740-50.

[3]In the empirical application of our model the Saudi Arabian crude petroleum price index was used as derived from the International Monetary Fund (IMF), *International Financial Statistics* (Washington, D.C.: IMF), various issues.

price of domestic currency in terms of the United States dollar.

Differentiating both sides of (2) with respect to time and then converting them into growth rates yields:

$$\dot{V}i/Vi = \sum_j bij\,\dot{X}ij/Xij + \sum_j bij\,\dot{P}j/Pj + \sum_j bij\,\dot{R}j/Rj \qquad (3)$$

where bij is the proportion of ith country's imports originating from the jth industrial country.

Equation (3) expresses the rate of growth of imports of any OPEC member country in terms of three items: (a) the rate of growth of the quantity of that country's imports; (b) the rate of growth of inflation in the industrial countries, and (c) a measure of the dollar's fluctuation against other currencies.

By integrating both sides of (3), we can write the total value of imports as the product of three terms, namely,

$$Vi = Ai \,.\, Pi \,.\, Xi \,.\, Ri$$

where

$$Pi = \int \sum_j bij\; Pj/Pj\; dt,$$

$$Xi = \int \sum_j bij\; Xij/Xijdt \text{ and}$$

$$Ri = \int \sum_j bij\; Rj/Rj\; dt$$

are the price, quantity, and exchange rate indices for ith country, and Ai is a constant term.[4]

From (3) it follows that the unit import price index in terms of United States dollar for ith OPEC member country is given by Pi Ri with its growth rate by,

$$\dot{P}mi/Pmi = \sum_j bij\,\dot{P}j/Pj + \sum_j bij\,\dot{R}j/Rj\,. \qquad (4)$$

By substituting (4) into (1.a) we arrive at the final equation which expresses the rate of change in ith country's terms of trade in terms of its determining factors, that is,

$$\dot{t}i/ti = \dot{P}o/Po - \sum_j bij\,\dot{P}j/Pj - \sum_j bij\,\dot{R}j/Rj\,. \qquad (5)$$

Equation (5) can be interpreted as a decomposition of the ith OPEC member country's terms of trade into three sources: (a) the rate of growth of the dollar price of oil; (b) a weighted average of the inflation rates in the

[4]It is important to note that the derivation of these indices is similar to the derivation of the Divisia quantity and price index in the theory of social accounting. See L.R. Christensen and D.W. Jorgenson, "The Measurement of U.S. Real Capital Input, 1929-1967," *Review of Income and Wealth*, December 1969 and "U.S. Real Product and Real Capital Input, 1929-1967," *Review of Income and Wealth*, March 1970. For this reason we call X, P, and R Divisia import quantity index, Divisia import price index, and Divisia exchange rate index, respectively.

industrial countries[5] and (c) a measure of the dollar's fluctuation from the viewpoint of ith country.

Similarly an equation can be derived to describe the sources of change in the terms of trade for OPEC as a whole. This is

$$\left(\frac{\dot{t}}{t}\right)\text{OPEC} = \frac{\dot{P}O}{Po} - \sum_j bj \frac{\dot{P}j}{Pj} - \sum_j bj\, \dot{R}j/Rj \qquad (6)$$

where in this case bj refers to the jth country's share in OPEC's total imports of goods and services.

The Empirical Results

OPEC as a Whole: In analyzing the sources of change of OPEC's terms of trade empirically we have considered the trade shares of OPEC with nine of the key industrial countries. These are: the United States, Japan, Belgium, France, Germany, Italy, the Netherlands, Switzerland, and the United Kingdom. Together these countries account for an overwhelmingly large proportion of OPEC's world trade. For instance, in 1977 they supplied about 76.7 percent of total OPEC's import needs.

The data on imports of OPEC as a whole and for individual countries were taken directly from the International Monetary Fund (IMF), *Direction of Trade*. These data refer to total imports of goods and services in millions of United States dollar. The data on exchange rates, export prices, and price of oil were taken from the IMF, *International Financial Statistics.* These include 29 quarterly observations from the fourth quarter of 1970 to the fourth quarter of 1977.

Table 1 reports the numerical results of our analysis of the sources of change in the terms of trade of OPEC as a whole for the time period 1971-1977. Perhaps the strongest impression conveyed by this table is the varying effect that the movement in the value of the dollar has had on OPEC's terms of trade. Over the period 1971-1973 where the dollar depreciated against most of the other industrial currencies, its downward movement eroded OPEC's terms of trade by 5.5 percent a year.

However, for the next three years, from 1974 through 1976, when the dollar recovered strongly from its previous low level and climbed noticeably vis-à-vis the British pound and Italian lira, its appreciation benefited OPEC

[5] Inflation here refers to the rates of change of export prices in the industrial countries. These are not clearly the most widely used indicators of inflation. Consumer prices or GDP deflators are the usual price indices on which inflation is measured. But because we are interested mainly in those aspects of inflation in the industrial countries which affect OPEC directly, we have opted for the export prices as the most appropriate price indices.

by increasing the implicit price of oil by about 2.3 percent a year. For the final year 1977, when the dollar set its downward trend once again, it cost the oil-exporting countries a 6.7 percent decline in the real export earnings. Averaged over the time period 1971-1977, the fluctuation in the value of the dollar has shrunk OPEC's purchasing power by about 2.3 percent a year. From the information in table 1 it can also be observed that inflation has been a more damaging factor in eroding OPEC's export earnings than has been the devaluation of the dollar. For instance, averaged over the whole period 1971-1977, the contribution of inflation amounted to about 9.03 percent as compared to about 2.3 percent due to dollar's fluctuation.

Table 1

THE SOURCES OF PERCENTAGE CHANGE IN OPEC'S TERMS OF TRADE[a]

Year	Term of Trade	Price of Oil	Inflation	Dollar's Fluctuation
1971	16.52	25.14	-3.99	-4.49
1972	3.908	8.66	-3.77	-2.51
1973	72.54	96.9	-14.83	-9.52
1974	106.52	130.46	-26.58	+2.63
1975	10.22	10.27	-1.69	+1.64
1976	-7.90	-.73	-9.93	+2.76
1977	-6.165	2.98	-2.43	-6.717
Annual average	27.80	39.097	-9.032	-2.31
Relative contribution	(100)	(140.63)	(-32.48)	(-8.30)

[a]OPEC members included in this calculation are Algeria, Indonesia, Iran, Iraq, Kuwait, Libya, Nigeria, Oman, Qatar, Saudi Arabia, United Arab Emirates, and Venezuela.

Individual Member Countries: Because OPEC members differ in terms of their trading and investment patterns, the impact of the dollar depreciation varies considerably from country to country. It is clear that nations such as Indonesia and Iraq, which purchase a large proportion of their import needs in markets outside the United States, bear a higher burden of foreign trade exchange losses than countries such as Iran, Saudi Arabia, Kuwait, and Venezuela.

Table 2 presents the contributions of the fluctuation in the value of the dollar and inflation to the rate of change of price of imports for these countries. Looking at this table we immediately realize that the devaluation of the dollar has increased the cost of imports to some countries much more than to others. For instance, the depreciation of the dollar during the time period 1971-1977 has contributed about 32 percent to the increase in the cost of imports of Indonesia, but it has contributed only 5 percent to the increase in the cost of Libyan imports. For each other country included in our sample and during the same period of time, the depreciation of the dollar has increased its import costs by about 20 percent a year.

Table 2

THE IMPACT OF THE DOLLAR'S FLUCTUATION AND INFLATION ON THE RATE OF CHANGE OF UNIT IMPORT PRICE OF A SELECTED NUMBER OF OPEC COUNTRIES, AVERAGED OVER 1971-1977[a]
(annual percentage rate of change)

Country	Dollar Fluctuation	Inflation	Import Price
Indonesia	3.74	7.61	11.35
	(32.95)	(67.05)	(100)
Iran	2.99	8.52	11.51
	(25.98)	(74.02)	(100)
Iraq	2.85	8.48	11.33
	(25.16)	(74.84)	(100)
Kuwait	2.52	8.89	11.41
	(22.08)	(77.92)	(100)
Libya	0.66	10.79	11.45
	(5.77)	(94.23)	(100)
Saudi Arabia	2.37	8.69	11.06
	(21.42)	(78.57)	(100)
Venezuela	1.86	8.85	10.71
	(17.36)	(82.27)	(100)

[a]Numbers appearing in columns in ( ) are relative contribution of each item to the rate of change of unit import price.

If we take into account the impact of dollar's fluctuation on the net external assets of these countries, the picture grows complicated; in this regard, however, a distinction has to be made between the countries which have a substantial proportion of their accumulated surplus or their outstanding debt in dollar-denominated assets and those nations which have achieved a more balanced diversification of their assets or debts across all currencies. Clearly, the losses or gains incurred as the result of dollar depreciation is more pronounced to OPEC members which have a higher degree of concentration of their assets or debts in terms of the dollar than the rest of the countries.

Table 3 presents some data on the net external assets of OPEC states for the years 1976 and 1977. As can be seen from this table, there are four countries, Algeria, Ecuador, Gabon, and Indonesia, having external liabilities which greatly exceed their external assets. These nations as a group had a net external liability of about $13 billion in 1976 and about $15 billion in 1977. Contrasting with these countries are four other countries — Kuwait, Saudi Arabia, Iran, and the United Arab Emirates — with net external assets accounting for about 90 percent of OPEC's $155 billion net external assets in 1977. Since a predominantly high proportion of these assets or debts are denominated in terms of the United States dollar, then any devaluation of the dollar will confer a substantial loss or gain to the surplus or deficit countries. It will hurt the surplus nations by eroding the purchasing power of their dollar holdings but at the same time it will benefit the deficit countries by reducing the real value of their dollar-denominated debts.

Table 3

NET EXTERNAL ASSETS OF OPEC COUNTRIES
(in billions of United States dollars)

Country	1976	1977
Kuwait	25	31
Qatar	4	5
United Arab Emirates	12	16
Saudi Arabia	56	120
Algeria	-3	-4
Ecuador	--	-1
Gabon	-1	-1
Indonesia	-9	-9
Iran	18	22
Iraq	.5	7
Libya	.6	8
Nigeria	.5	3
Venezuela	9	10
Total	126	155

Source: Morgan Guaranty Trust Company, *World Financial Markets* (New York), November 1977.

The wide disparity observed in the experience of different OPEC members in bearing the cost of the dollar's depreciation is both interesting and important. It is important because it brings to the surface some of the potential conflicts in OPEC states associated with the existing dollar oil-pricing system and, simultaneously, it points to the serious problems facing OPEC in attempting to change its pricing mechanism from a dollar-based system to any other currency system.

These potential conflicts stem partly from the wide differences in trade patterns existing among OPEC members and partly from the erratic behavior of industrial currencies. It is undoubtedly true that as far as there are differences in the trade directions of OPEC countries and as far as there is no uniformity in the movements of industrial currencies, different countries will be subjected to different cost of exchange rate changes. It is also true that the cost will be higher to nations whose import shares are weighted mostly toward the strong-currency countries than nations whose imports originated mainly from the weak-currency countries. This, in fact, seems to be the main reason underlying the difference observed between Indonesia and Libya in terms of the increase in their cost of imports attributable to the dollar depreciation. Indonesia's principle supplier of imports during the 1971-1977 span was Japan[6] with a currency which has appreciated tremendously against the dollar, whereas Libya's main trading partner was Italy[7] with a currency which has lost value appreciably against the dollar.

Conclusion

In this paper the sources of growth in OPEC's terms of trade over the period 1971-1977 have been identified, analyzed, and their contributions were quantified. These are: (a) the growth in the dollar price of oil, (b) inflation in OPEC's trading partners, and (c) the fluctuation of the dollar against other currencies. Over the whole period it was found that the fluctuation in the value of the dollar contributed by about 16.24 percent and inflation by about 63.2 percent to the decline in OPEC's terms of trade.

Turning to the impact of the dollar's fluctuation on individual OPEC members, it was determined that Indonesia was the hardest hit and Libya was the least hurt by the depreciation of the dollar over the period 1971-1977. The experience of other member countries was more or less uniform. This difference in the experience of OPEC states was then attributed to the difference in their trade structure and to the nonuniform behavior of the major currencies against each other.

---

[6]During the period 1971-1977, Japan on average supplied about 45 percent of Indonesia's total import requirements.

[7]During the period 1971-1977, Italy's share of Libya's imports amounted to about 34.6 percent.

17

OPEC'S ROLE IN A GLOBAL ENERGY AND DEVELOPMENT CONFERENCE

*Herbert E. Hansen**

The Need for OPEC Sponsorship in International Negotiations on Energy and Development

In the first few days ushering in the decade of the 1980s, nothing is more important than to develop a workable scenario for a meaningful producer-consumer dialogue on energy and development. A snapshot of the political and economic condition of the Middle East world at the beginning of the '80s seems only to have slightly postponed Paul Erdman's scenario for the *Crash of '79:*

1. The lack of stability in Iran as well as the embargo and complete breach with the United States over U.S. embassy hostages.
2. The resurgence of the Shi'ite religion in Iran and of Islamic movements generally must appear as a backlash against the West's policies in various Moslem countries, although these have often been utilized for other purposes by other interests.
3. The invasion of Afghanistan by Russia, and the Soviet Union's increasingly disruptive influence or attempts at influence in such countries as Iraq, Syria, North and South Yemen, Ethiopia, and potentially Pakistan and India (all nations circling the more conservative Arab oil-producing states and therefore a perceived threat to their existence).

---

*Herbert E. Hansen, Vice President for Government Agreements of Gulf Oil Exploration and Production Company in Houston, earned a M.B.A. from Harvard Business School and a doctoral degree in law from the same university. He has held positions in Gulf's offices and affiliates in Tehran, London, and Pittsburgh. The author's contributions to the 1975 international energy conference at the University of Colorado appeared in the spring 1976 issue of the *Journal of Energy and Development*. His participation in subsequent conferences appears in the published proceedings, including *New Policy Imperatives for Energy Producers* (Boulder, Colorado: International Research Center for Energy and Economic Development, 1980).

4. The existing chaos in oil production levels, offtake restrictions and prices, which make impossible economic planning and development in both oil-consuming and producing countries.

5. The disintegration of the established role of international law as the "have-nots" press growing demands on the "haves." All of this paints a bleak picture for energy, economics, and politics today. The basic political risk we face is one of international economic chaos and military survival, with the "have-nots" demanding a great deal more from the "haves" and, in the process, largely ignoring established legal conventions. The doctrine of changing circumstances has truly replaced the doctrine of contractual stability in most areas. And our response has been inadequate. The political situation at the beginning of the 1980s reminds me of Winston Churchill's statement to Parliament in 1941 when Rudolph Hess parachuted into Scotland: "This is one of those cases in which the imagination is baffled by the facts."

6. If we look at the world's power blocs and international organizations which have been or might become involved in such a dialogue, the OPEC countries may be the one group that effectively could move us towards the establishment of a new economic order within the required time frame to prevent disaster. The United States is currently handicapped by its foreign policies and its confrontation attitude towards OPEC. The United Nations, its affiliates and specialized agencies as UNCTAD, and other multinational organizations as the IEA, OECD, and even the positive Paris meetings of a much smaller CIEC, do not seem able to construct the organization or forum for involving OPEC, consumers, producers, and LDCs in a constructive way.[1]

On the other hand, the member countries of OPEC (acting through its Secretariat if desired) could reap extensive benefits by adopting a responsi-

---

[1]The acronyms and abbreviations used in this paper include: OPEC, Organization of the Petroleum Exporting Countries, composed of 13 member states (Algeria, Ecuador, Gabon, Indonesia, Iran, Iraq, Kuwait, Libya, Nigeria, Qatar, Saudi Arabia, United Arab Emirates, and Venezuela) and headquartered in Vienna; UNCTAD, United Nations Conference on Trade and Development (Geneva); OECD, Paris-headquartered Organization for Economic Cooperation and Development is composed of Australia, Austria, Belgium, Canada, Denmark, Finland, France, Federal Republic of Germany, Greece, Iceland, Ireland, Italy, Japan, Luxembourg, Netherlands, New Zealand, Norway, Portugal, Spain, Sweden, Switzerland, Turkey, United Kingdom, and United States; IEA or International Energy Agency of the OECD included participation as full or associate members those countries belonging to the OECD except Australia, Finland, France, Iceland, and Portugal; CIEC, Conference on International Economic Cooperation often called the North-South Dialogue, initiated in the mid-1970s; LDCs or less-developed countries; IMF, International Monetary Fund; GATT, General Agreement on Tariffs and Trade ; SDR or Special Drawing Right as monetary unit employed by the IMF to circumvent the problems of fluctuating values of various national currencies.

ble, statesmanlike role in calling for a global conference on energy and development under the sponsorship of an independent agency such as the World Bank. In this event, it seems clear that the consuming countries would have no alternative but to respond positively and to participate willingly. Not only would such an OPEC "coup" establish OPEC as a mature and responsible political force in international affairs, but it might serve notice that the body could become the leader needed to foster progress and growth worldwide in an orderly fashion. OPEC would also reap additional benefits from the successful conclusion of such a conference. For example, it could bring about conditions conducive to stable investment conditions in the industrialized world, predictable national income policies and development programs for all countries, a moderation of rampant global inflation, restraint of excessive speculation in oil spot markets, and plans for assistance to developing countries, to name a few.

Organizations Currently Involved in North-South Dialogue

First, let us review the many and varied initiatives working towards such a cooperative dialogue presently under way. They include the following:

1. The United Nations Committee of the Whole involves the current United Nations global negotiations on the North-South Dialogue, which will include all 153 members of that body, haggling over everything from the international financial system to commodity prices, aid to developing countries, trade barriers, and the supply and price of oil and other energy forms in the most political forum of all forums. Historically this has not proved productive, despite the constructive initiative in 1975 and 1977 to shape a Conference on International Economic Cooperation (CIEC) in Paris limited to 27 major nations of the industrial bloc, large oil-producing countries, and the Third World. It failed because no one was prepared to give up anything of importance at that time.

The United Nations Committee of the Whole is drawing up a detailed agenda, to be followed by a Special Session of the General Assembly about September 1980, which in turn will be followed by what in all probability will result in prolonged political polemics, even less productive than a 10-year Conference on the Law of the Sea. While the United Nations initiatives cannot be ignored because this is truly a global problem, in my opinion there *must* be a more rewarding forum for such discussions. Again, to quote Churchill apropos of the marriage of his American and English parents, "There is no plant in the whole world of more cautious growth than Anglo-American negotiation."

2. The Rockefeller Foundation's "Group of 30" chaired by Dr. Witteveen (formerly head of the IMF), is formally known as the "Consultative

Group on International Economic and Monetary Affairs'' and made up of 30 leading bankers, officials and businessmen from around the world.

3. The EEC-Arab dialogue reflects initiatives undertaken toward dialogue between the European Economic Community (EEC) and the Arab oil-producing countries, and between the OECD and OPEC nations. To date, none of these have produced any specific progress.

4. A new development will involve the National Security Council's Interagency Task Force on ''Petroleum Exploration in the Oil-Importing Developing Countries,'' whose draft report is scheduled to be finalized this month (March). It is chaired by Mr. Ed Fried, a former United States representative to the World Bank, and consists of representatives from all other United States government agencies involved. Briefly, an earlier draft proposed participation by the United States government in LDC-oil company negotiations and the potential establishment of an Exploration Revolving Fund to be administered bilaterally by the United States (or possibly by a multilateral organization.) The stated objective would be to reduce financial risk for developing countries and to stimulate oil exploration, including staffing of exploratory drilling in high-risk areas because they view the World Bank and industry efforts in those countries as inadequate and too high priced for them to handle. The dangerous political implications and the complete lack of practical and technical capabilities inherent in a United States national oil company approach as an independent government program are obvious. So, too, are the erroneous assumptions regarding the World Bank and oil industry efforts upon which it is based. Hopefully, it will be modified substantially before becoming U.S. policy.

5. World Bank program for oil and gas exploration and development is a new undertaking which this author has personally used and helped sponsor. I believe it will successfully and realistically accomplish much in both exploration and development in LDCs in conjunction with the oil companies. In the process, it will involve important political risk protections for both the developing country and the oil operators.

6. Another constructive influence throughout has been the multitudinous seminars, conferences, and studies sponsored by universities, foundations, and research institutes on energy and development over the past seven years. To mention a few: The annual international energy conference sponsored by the International Research Center for Energy and Economic Development and held at the University of Colorado at Boulder; energy and international seminars sponsored by the Fletcher School of Law and Diplomacy and by Harvard University; similar sessions sponsored by Georgetown University and the new Program for International Business Diplomacy headquartered there; the East-West Center in Honolulu; the Oxford Energy Seminar at Oxford University in England; OPEC's own energy seminars held at least annually in Vienna; the recently formed and innovative

Center for International Business, headquartered in Houston and Dallas; and scores of other major institutions, foundations, and research organizations. There is an old saw that says "If you laid all our economists end to end, it would be a good thing," but I do not subscribe to this view entirely, due in part to such positive activities just noted.

7. The World Energy Conference is scheduled for Munich in September 1980, with 4,000 attending from among its 70 national committee representatives and guests.

8. The Society for International Development's Energy Roundtable for the North-South Dialogue (in Rome and Washington) has undertaken a three-year multimillion dollar study that seems both realistic and constructive. But are they consulting with the industry in a meaningful manner?

9. The Willy Brandt Commission of 21 also known as the Independent Commission on International Development Issues consists of a very high level 21-member commission headed by ex-Chancellor Willy Brandt. It has submitted a three-year study "North-South: A Program for Survival" to the United Nations Secretary General. The report contains many interesting and novel suggestions, including an international tax levy to support a new world development fund designed to act as a super-welfare agency which would dole out aid without regard to specific or productive projects.

10. The United Nations Conference on New and Renewable Sources of Energy, the Law of the Sea Conference, and some 5,998 other conferences and meetings annually form a major body of initiatives in this area. The Conference on New and Renewable Sources of Energy will meet in Nairobi, Kenya, in August 1981, with eight panels chaired by Japan's Taniguchi. Indeed, the Law of the Sea Conference itself can be considered as part of the United Nations North-South Dialogue. Each year there are over 6,000 United Nations conferences involving more than one million pages of documentation.

11. Other initiatives and proposals are reported daily by the media in such publications as the *Wall Street Journal, Platt's Oilgram, Middle East Economic Survey, Petroleum Intelligence Weekly* on how we should go about this all-important dialogue. The above listing is in no way offered as a comprehensive one.

Action Options

What can we do about all these conflicting countercurrents based upon the perceptions of their own self-interest by each of the involved groups? Let me suggest four major options for consideration.

First, let present circumstances run their course. As Confucius said in his *Analects,* "The cautious seldom err." However, I must confess in my view doing nothing in today's circumstances could only result in complete disaster. To quote Gulf's Chairman, Dr. Jerry McAfee, "We have been rearranging the deck chairs on the Titanic," rather than biting the bullet.

Second, confront OPEC and the producing countries with a consumer bloc designed to break their control over conventional oil and gas energy production and pricing. To some degree this has been the unfortunate United States policy during most of the past 10 years, and it certainly has not been productive for the simple reason that there are *no enforceable economic limits at the present time* to production levels and prices of oil other than alternative sources of energy. Those of us intimately involved in the trade fully understand the basic precept of energy, that we will continue to be dependent primarily on oil and gas production and upon approximately the same level of imports of crude oil from OPEC Middle East countries to the United States during the 1980s.

Third, a variation of the military intervention option might lead one to defensive or even offensive military intervention in the Middle East or elsewhere to protect our access to global energy supplies. There is not time today to discuss the possible consequences of this option, but I believe that in addition to running a very high risk of self-destruction, there does exist a *practical alternative.*

The fourth option is of cooperation between producer/consumer countries in an international dialogue. This approach should be tabled now for immediate action, e.g., cooperation between producing and consuming countries in an international dialogue. From the standpoint of an oil man, it is clear that conventional energy supplies are limited and that their allocation is a global resource, a resource in which we all must share. We are *interdependent* whether we like it or not. Any satisfactory solution can only be arrived at by consultation and negotiation.

An OPEC Solution

The most practical solution to bring about cooperation between producers and consumers in my opinion is for the OPEC countries as an organization to take the lead in world statesmanship by calling for a global conference now, and for them to suggest that an independent outside agency such as the World Bank act as sponsor and catalyst in setting up a practical working-level conference. Obviously this approach involves many difficult questions. For example: How can the industry input be successfully brought in? In the past, too little advice and consultation from industry and those with the knowledge and ability to solve the problems has been requested by the governments. As Prime Minister Clemenceau said in 1886: "War is much too serious a matter to be entrusted to the military." So, too, it must be with global energy and our worldwide politicians without industry's participation. In fact, OPEC has already devoted much time and attention to such a dialogue as evidenced by the statements of many of the oil ministers and by their participation in such international energy conferences noted earlier in Colorado, Oxford, and their own OPEC Seminars in Vienna.

In May 1978, OPEC set up a long-term strategy committee in Taif, Saudi Arabia, to examine OPEC's policy options for the 1980s. Members of the working group under this committee are 'Abd al-Amir Anbari (Iraq), Fadhil al-Chalabi (OPEC), Farouk al-Husseini (Saudi Arabia), Francisco Parra (Venezuela), Nordine Ait-Laoussine (Algeria), Nureddin Farrag (Kuwait), and Parvis Mina (Iran) reporting to their ministers chaired by Sheikh Yamani of Saudi Arabia. Their report was finalized in December 1979 and tabled at the OPEC Caracas conference. The report contains five chapters, including one on the basis for a broad dialogue with industrialized nations aimed at establishing some stability in oil supplies; the other four chapter heads encompassed oil and energy perspectives over the next 10 years, relations with developing countries, relations with the industrial world, and a long-term "rational" oil-pricing system.

Agenda Problems and Some Suggestions

If OPEC is to come to the table at all, the conference must also involve developing countries (including issues of aid, energy exploration and financing), possible changes in the international monetary system, and other major initiatives.

1. What role for the "have-nots" (which incidentally actually includes some of the OPEC countries). The shaping of the agenda in and of itself is a major undertaking, but it can be done, and international agencies such as the IMF, GATT, and UNCTAD can be enlisted to help. Eventually, such a conference could report back to the United Nations Committee of the Whole, which is already started up and in motion on this problem.

2. Should there be a role in these negotiations for China and/or Russia? To date no one has seriously undertaken to involve either one in preliminary planning, although their role in world energy is obviously important, and they would participate in the United Nations Conference as members.

3. What role for the World Bank? No other independent international agency has the experience, financial resources, personnel, and credentials with the LDCs, OPEC, and industrialized countries to match. It should act as cosponsor, technical advisor, and coordinating agency to bring all the multifaceted strains of energy and development necessary to put together a smaller working group, geared to find solutions for eventual presentation to the United Nations Committee of the Whole.

The World Bank is already deeply involved in oil, gas and other energy exploration and development projects on a large scale, in almost all cases utilizing the international industry and national companies with the required expertise. For example, the current program through June 1982, involves 44 operations leading to loans and credits of $1.36 billion in oil and gas, $1.96 billion for electric power development by June 1980, and 80 coal projects in 7 member countries will lead to Bank lending of $600 million. The Bank also assists 55 member countries in developing national energy plans and policies, including technical advice and negotiation training, with emphasis on the non-oil-exporting LDCs. In the process they have developed a trained cadre of engineers, economists, political analysts, and energy executives.

Furthermore, the Bank's lending is but the tip of the iceberg since it acts only as a catalyst, encouraging other financial sources to become involved and offering a degree of political risk protection to *all* parties not otherwise available. Take the Gulf of Siam pipeline for Thailand as an example. A $5 million Bank loan for engineering led to a $107 million Bank loan which will assist in planning a gas development project involving, in three states, a total investment of $1.8 billion, of which $1 billion will be made by foreign oil companies and their commercial bankers, and $800 million by the Government of Thailand and its various sources of finance, including private bank groups.

The Bank's authorized capital base recently has been doubled to $80 billion, and its related agencies, the International Finance Corporation (commercial investments), the International Development Association (soft or low-interest loans), and the separate but interrelated International Monetary Fund, can and will play the leading role in international finance and development structuring. Perhaps the OPEC "Development Agency," formerly the OPEC Special Fund, which is planning to increase its assets to $4 billion for LDC aid, could coordinate its activities with the World Bank group on productive energy projects for developing countries worldwide. Certainly those nations look naturally to the World Bank to assist them in economic survival. Since a basic political risk inhibiting the further use of private capital investment is the growing disregard for international law by the "have-nots" grasping for more from the "haves," this would seem to offer an opportunity to secure private investment in countries now deemed as being too high political and economic risks. Our company (Gulf) has utilized the Bank services in Pakistan and other nations and have found it to be creative and useful in establishing risk protection for *all* parties concerned. The ultimate political risk protection can only be arrived at by a cooperative effort involving all interests. As Woodrow Wilson addressed the United States Senate in January 1917: "There must be, not a balance of power, but a community of power; not organized rivalries, but an organized common peace." Thus the World Bank is uniquely fitted for its leading role as an independent but deeply involved international agency.[2]

The agenda for such a conference will be both complex and controversial. While the West wishes to limit it to oil prices and stable supplies, OPEC believes that the basis for a new international economic order must be established, including reform of the monetary system, aid to developing countries, and related trade and development agreements. Many of the possible items for discussion have been thoroughly spelled out in Øystein Noreng's recent book *Oil Politics in the 1980s*, which is certainly worthy of much discussion.[3] I suggest that a workable agenda to be proposed by OPEC might include the following issues.

4. The need to match production of energy overall to consumption is real. Since oil and gas, in particular, and energy, generally, are finite global assets, they must be allocated to the benefit of all nations in the long term. Therefore, it would be logical for the conference to try to match oil and gas production to consumption. Although consumption of oil and gas as the primary energy might decline, an overall stability can be envisioned as a result of more energy-efficient uses resulting from conservation and the gradual development of alternative forms of energy to displace oil and gas over the longer term. Incentives could be agreed upon to foster increased exploration in OPEC states for oil and gas (and other natural resources) if OPEC could be reassured that oil and gas produced would be as valuable in the form of international investment as if they were left in the ground. Negotiations regarding the production and processing of natural gas liquids, which currently constitute as much as one-half of the world's hydrocarbon Btu's, should also be included in achieving an overall supply/demand balance.

5. Trade and international finance problems, including those of developing nations, could be included in the agenda and then most profitably referred to working groups in established international agencies such as the IMF, GATT and UNCTAD, to report back to the conference.

6. Joint ventures between producing and consuming countries, utilizing the necessary capabilities and expertise of international companies, could be proposed for investment in alternative energy projects to the mutual benefit of all concerned.

---

[2]The World Bank has an active publishing and information program to which the researcher may turn — annual reports, bulletins, country studies and the like. The International Monetary Fund issues statistical data in such publications as *Directions of Trade* and the *IMF Survey*.

[3]Øystein Noreng, *Oil Politics in the 1980s* (New York: McGraw-Hill Book Company, 1978).

7. A program to stimulate energy exploration and production in developing countries, such as that currently being implemented by the World Bank, the UN Development Program, and other organizations (in cooperation with the international companies), would not only assist the economic balance of those countries, but also permit the use of agency funds, otherwise directed to unproductive uses, to be applied to development programs in those nations.

8. A pricing mechanism for oil and gas and an oil futures market, which would permit predictable price increases tied to an agreed formula, would have to become part of the negotiated package if satisfactory results are to be obtained. We believe that, although OPEC maintains that price policies are their exclusive and sovereign province, they might see the merits of a mutually acceptable formula which they could voluntarily adopt. Many such proposals have been tabled and discussed in the past, including the indexing of oil prices to automatic inflation rates plus a percentage, to a basket of currencies expressed by SDRs or Special Drawing Rights, to dollar value fluctuations against nine other currencies (as was contained in the so-called Geneva agreement between the oil industry and the OPEC countries of January 20, 1972), and even recently a system patterned upon that applicable to international commodity agreements in copper, sugar, and tin. Under these agreements there is a negotiated minimum and maximum price, between which the free market operates until intervention becomes necessary when stocks are drawn down from stockpiles. Although it would not seem advisable to physically run down strategic oil stockpiles, perhaps it would be possible to develop *an oil futures market* which would permit individual companies or institutions to buy and sell crude oil and products futures like other commodities, thereby hedging against excessive price fluctuations.

In fact, there already exists such a market administered by the New York Mercantile Exchange which has dealt in contracts for #2 (home-heating and diesel oil) and #6 (industrial) oil since November 1978. Trading volume is significant and rising, nearing 1,000 contracts for 1,000 barrels daily, including trading now beginning in regular and unleaded gasoline. Actual delivery has run only 5 to 10 percent, since it is mainly used by speculators and hedgers, thereby moderating the wilder swings in the market.

Since this is a new market, many restrictions are necessary — time and place of delivery, specifications for quality control, fines for defaulters or violators, a variable daily limit for price changes as in other commodities, to name a few. The risks of weather, disasters, inventory changes, supply/demand policies, worldwide political events, and storage limitations make this a market primarily used by the refiners, jobbers, distributors and retailers of large quantities of products who are hedging their positions for a budget year. This is done principally against the broad swings in the spot market, which is dominated by about 100 independents and oil traders who are in contact on a daily basis.[4]

9. Timing is critical. Anything as complicated as this proposed conference will indeed take time, but we must start now in order to arrive at a satisfactory solution before the lack of results overwhelms us. Economic summit meetings of the industrialized nations are scheduled for June 22-23, 1980, in Venice, Italy, and OPEC also has tentatively scheduled a summit meeting, probably in Baghdad in October of 1980. The interim period might well be utilized productively by special study groups set up with the assistance of foundations and university programs as well as by industry advisory groups to develop various policy options for later consideration by the conference.

United States Policies Required

How then, can we in the United States help to accelerate the consultative process?

1. We must recognize that a *meaningful dialogue* with OPEC and the oil producers is both necessary and desirable, and that it cannot be limited to energy alone.

2. We must help change the attitude of the Congress, *government agencies, and media in the United States* from a search for "scapegoatism," to one of willingness to publicize and act on creative and cooperative approaches.

3. *United States Foreign policy* must be based upon the United States self-interest worldwide, and in my opinion the Soviet Union has given us a great assist in this regard by crudely and openly declaring their expansionist objectives in Afghanistan. But until we openly recognize that the military and oil security in the Middle East are indisputably linked to a solution of the Palestinian settlements in the West Bank, Gaza, and Jerusalem, it will continue to poison all of the relations with all Arab countries involved in such a dialogue.

4. United States government and business interests must get closer together by consultation and direct advisory meetings. There has been a notable *increase in this cooperation* within the last six months by frequent agency consultations with industry leaders.

5. There exists a need to reduce excessive overregulation and interference in the free market system by governments in those areas where monopoly is clearly not an economic consideration.

---

[4]The *Houston Business Journal*, January 28, 1980, reported on this market.

Conclusion

I cannot resist a final quotation from Henry Kissinger's mighty tome of psychological, historical, and political wisdom in the *White House Years*.

> History knows no resting places and no plateaus. All societies of which history informs us went through periods of decline; most of them eventually collapsed. Yet there is a margin between necessity and accident, in which the statesman by perseverance and intuition must choose and thereby shape the destiny of his people.[5]

Many reading this article can help; hopefully, enough will be willing to work together towards resolving this dilemma. What is needed *now* is to trigger these actions by an urgent call from OPEC to the consuming nations and to the LDCs to cooperate in organizing a workable, realistic conference sponsored and assisted by the World Bank.

[5]Henry Kissinger, *The White House Years* (New York: Little, Brown, 1979).

INDEX

For Product Safety Concerns and Information please contact our EU
representative GPSR@taylorandfrancis.com
Taylor & Francis Verlag GmbH, Kaufingerstraße 24, 80331 München, Germany

www.ingramcontent.com/pod-product-compliance
Ingram Content Group UK Ltd.
Pitfield, Milton Keynes, MK11 3LW, UK
UKHW020015200926
475702UK00014B/1900
* 9 7 8 1 1 3 8 6 4 2 8 9 8 *